Against the Tides

The Nature | History | Society series is devoted to the publication of high-quality scholarship in environmental history and allied fields. Its broad compass is signalled by its title: *nature* because it takes the natural world seriously; *history* because it aims to foster work that has temporal depth; and *society* because its essential concern is with the interface between nature and society, broadly conceived. The series is avowedly interdisciplinary and is open to the work of anthropologists, ecologists, historians, geographers, literary scholars, political scientists, sociologists, and others whose interests resonate with its mandate. It offers a timely outlet for lively, innovative, and well-written work on the interaction of people and nature through time in North America.

General Editor: Graeme Wynn, University of British Columbia

A list of titles in the series appears at the end of the book.

NATURE | HISTORY | SOCIETY

Against the Tides

Reshaping Landscape and Community in Canada's Maritime Marshlands

RONALD RUDIN

FOREWORD BY GRAEME WYNN

© UBC Press 2021

All rights reserved. No part of this publication may be reproduced, stored in a retrieval system, or transmitted, in any form or by any means, without prior written permission of the publisher, or, in Canada, in the case of photocopying or other reprographic copying, a licence from Access Copyright, www.accesscopyright.ca.

30 29 28 27 26 25 24 23 22 21   5 4 3 2 1

Printed in Canada on FSC-certified ancient-forest-free paper (100% post-consumer recycled) that is processed chlorine- and acid-free.

Library and Archives Canada Cataloguing in Publication

Title: Against the tides : reshaping landscape and community in Canada's maritime marshlands / Ronald Rudin.
Names: Rudin, Ronald, author.
Series: Nature, history, society.
Description: Series statement: Nature, history, society | Includes bibliographical references.
Identifiers: Canadiana (print) 20210216700 | Canadiana (ebook) 20210217847 | ISBN 9780774866750 (hardcover) | ISBN 9780774866774 (PDF) | ISBN 9780774866781 (EPUB)
Subjects: LCSH: Canada. Maritime Marshland Rehabilitation Administration – History. | LCSH: Reclamation of land – Fundy, Bay of, Region – History. | LCSH: Flood dams and reservoirs – Fundy, Bay of, Region – History. | LCSH: Salt marshes – Fundy, Bay of, Region – History. | LCSH: Fundy, Bay of, Region – History.
Classification: LCC TC978.C32 F86 2021 | DDC 627/.540916345 – dc23

Canadä

UBC Press gratefully acknowledges the financial support for our publishing program of the Government of Canada (through the Canada Book Fund), the Canada Council for the Arts, and the British Columbia Arts Council.

This book has been published with the help of a grant from the Canadian Federation for the Humanities and Social Sciences, through the Awards to Scholarly Publications Program, using funds provided by the Social Sciences and Humanities Research Council of Canada.

UBC Press
The University of British Columbia
2029 West Mall
Vancouver, BC V6T 1Z2
www.ubcpress.ca

To Ernie Partridge

Contents

List of Illustrations / ix

Foreword: Poets among the Engineers / xi
by Graeme Wynn

Preface: In Search of the MMRA / xxv

A Note on Sources / xxxii

Prologue: Whose Knowledge? Which Nature? / 3

Part 1: Second Nature

1 Out to Sea / 25

2 Reconstruction / 72

Part 2: Third Nature

3 Dam Projects / 117

4 Legacies / 170

Epilogue: Meet the Grand Pre Marsh Body / 220

Notes / 226

Bibliography / 265

Index / 276

Illustrations

Maps

1 Marshlands in Nova Scotia and New Brunswick / 5
2 MMRA tidal dams / 120

Figures

0.1 Trying to hold back the tides in *Les aboiteaux* / 7
1.1 Middle Dyke Road flood, Grand-Pré, Nova Scotia, mid-1940s / 27
1.2 Uplands and marshlands / 31
1.3 Cross-section of dykelands / 33
1.4 Drainage ditches expel water from field / 34
1.5 Ernie Partridge and his dyking spades / 36
1.6 Grand Pre Marsh Body commemorative plaque / 41
1.7 Work crew on the Wallace Bay aboiteau, ca. 1910 / 44
1.8 Hay barn on the Tantramar Marsh, New Brunswick / 44
1.9 Dragline at John Lusby Marsh, early 1950s / 49
1.10 Repairing aboiteau after tiding, Falmouth Village Great Dyke Marsh Body, 1906 / 52
2.1 Rebuilding an aboiteau in *Les aboiteaux* / 75

2.2   Canada Car and Foundry building, Amherst, NS, 1931 / 78
2.3   Prefabricated lumber for sluices, MMRA workshop, Amherst, NS / 80
2.4   Excavating sluice bed, Comeau Marsh, July 1949 / 86
2.5   Assembling prefabricated aboiteau, Grand Pre Marsh Body, 1950 / 86
2.6   Bishop Beckwith Marsh Body, MMRA survey map, NS 65, 1957 / 91
3.1–3.4   Construction of Shepody Dam / 128–29
3.5   Hayfield in Tantramar Marsh / 133
3.6   Annapolis River Dam work site / 143
3.7   Annapolis River Dam nearing completion / 143
3.8   Postcard of Petitcodiac tidal bore / 152
3.9   Surfers on the Petitcodiac, 1967 / 152
3.10–3.11   Petitcodiac River at Moncton, 1954 and 1967 / 160–61
3.12   Watching the closure of the Avon River, June 12, 1970 / 168
4.1   Surfers on the Petitcodiac, July 2013 / 172
4.2   Flooded hay field, near Sackville, NB, 1951 / 176
4.3   Aerial photo, Lake Petitcodiac, 1995 / 180
4.4–4.5   Petitcodiac River at Moncton, 1963 and 1976 / 184–85
4.6   Student art against the Petitcodiac Causeway / 195
4.7   Robert F. Kennedy Jr. at Moncton, along the Petitcodiac River, with Julia Chadwick and Michel DesNeiges, February 24, 1995 / 201
4.8   Ramp leading to the former marina on Lake Petitcodiac, Riverview, NB / 207
4.9   Petitcodiac salt marsh downstream from Moncton, 2003 / 209
4.10–4.12   Accumulation of silt downstream from Avon Causeway / 210–11
4.13   Windsor Salt Marsh / 213
4.14   Save Lake Pesaquid campaign / 216
4.15   Bass fishing tournament on Lake Petitcodiac, pre-2010 / 218
4.16   Lake Pesaquid Pumpkin Regatta, 2012 / 218
EP.1   Signage posted by Grand Pre Marsh Body / 223

FOREWORD

Poets among the Engineers

Graeme Wynn

AGAINST THE TIDES shines an illuminating spotlight on the work of Canada's Maritime Marshland Rehabilitation Administration (MMRA), a mid-twentieth-century government body that substantially remade some of the most striking, storied, and oldest European-settled landscapes in northern North America. The book focuses on the Bay of Fundy region during the quarter century after the Second World War to reflect on the oft-manifest tension between local knowledge and scientific expertise.

By drawing on previously neglected archival records pertaining to the administration of the marshes, on conversations with many of those involved in or affected by the work of the MMRA, and on broader scholarly literatures, Rudin succeeds in making new sense of a surprisingly tangled, fragmented, even chaotic, story of people, politics, ecologies, and landscapes. Along the way, he also reveals his deep familiarity with the places about which he writes and the individuals who lived and worked in them – and, unusually, he invites readers to share this awareness by viewing *Unnatural Landscapes*, a short film made in conjunction with the book.[1]

I have considerable personal affection for the people and places at the centre of this story. My first Canadian research began with the marsh mud that is central to the history of the Bay of Fundy, and my first visits to this area occurred in the summer of 1969, just as the Maritime Marshland Rehabilitation Administration was wrapping up the two decades of transformative endeavour detailed in the pages that follow. On the threshold of my graduate training, I tried my darnedest to meet the expectations implicit in my desire to become an historical geographer. In regional archives, libraries, museums, and the backrooms of local historical societies I did the historian thing, spending weeks poring over old documents and manuscript maps. Most weekends, I laced up my geographer's boots and walked the roads, dykes, and farm tracks of Grand-Pré, the Annapolis Basin, the Avon and Cornwallis Rivers, and Chignecto at the very head of the great tidal bay.[2]

My research gaze was fixed resolutely on the past, a past some two centuries distant, because I was intent on learning what I could of how the marshes were used in the half-century or so after the Acadian expulsion from these lands in 1755. The Acadian achievement in reclaiming productive fields from the onrushing tides had been much remarked, celebrated, and documented. After halting beginnings, when the Habitation established by Samuel de Champlain and a small band of adventurers at Port-Royal on the Dauphin (Annapolis) River in 1605 was destroyed by raiders from Virginia, French settlers returned to the area – occupied for millennia by Mi'kmaq – in the 1630s. By mid-century, two to three hundred newcomers had turned tidal marshes (meadows) along the river into productive fields. Two decades later "a settled and thoroughly acclimated people" – some 67 families or 350 Acadians – were established there.[3] By the end of the seventeenth century. there were several offshoot settlements around the Basin of Minas and Cumberland Basin. In the summer of 1755, when the British began to remove Acadians from their lands (a campaign known as the Deportation or *le grand dérangement*), there were probably about 9,000 of them in the Bay of Fundy settlements; perhaps a third as many had relocated elsewhere in the tumultuous years that preceded the deportation.

By standard accounts, the land – possibly 20,000 acres or 8,000 hectares of it – that Acadians had reclaimed from the Fundy tides lay abandoned after 1755. Dykes and the sluices (aboiteaux) that drained the reclaimed land fell into disrepair, and then, in November 1759, high winds and tides breached the ramparts.[4] The New England settlers (known as "Planters" to historians of Nova Scotia), who came to the Minas Basin and the Chignecto isthmus after 1760, "lacked the skill and labour to redeem the dyked marshlands." Or so thought the Canadian-American historian J.B. Brebner and many others.[5] In the manner expected of a graduate student, I was dubious. This account seemed altogether too neat and tidy. My summer of archival research brought to light just enough evidence to suggest that those early Planters were less impractical and unimaginative than Brebner had implied, and that they turned, quickly, to repair and maintain dykes and fields. Happily, this conclusion was confirmed, in later years, by other researchers.[6]

As I recall, my doubts about prevailing views of dykeland agriculture in the 1760s stemmed in the first instance from the work of new social historians, emphasizing the agency of ordinary men and women in shaping their lives.[7] My second seed of doubt grew from a sense that the cataclysmic deportation had encouraged a romantic view of pre-Expulsion Acadian society that exaggerated the uniqueness and magnitude of their land reclamation efforts.[8] Much was made, in discussion of these achievements, of the enormity of the Fundy tides. But my weekend walks were enough to sow doubts about this too. Especially on Grand-Pré, I was puzzled by what I thought of as "inland dykes," remnants, usually broken and incomplete, of structures far from the current outer edge of the marsh. I inferred, tentatively, that reclamation had probably been done in stages, beginning close to the upland edge of the original salt marsh, where the diurnal tidal range would have been measured in centimetres, rather than the metres usually ascribed to the waters of the bay, and converting, piecemeal, tens of acres at a time, rather than the many hundreds that surrounded me. I had neither the time nor the wit to solve this puzzle completely. Again happily, Sherman Bleakney did so brilliantly and with far more insight and authority than I could possibly muster, in his remarkable book *Sods, Soils and Spades: The Acadians of Grand Pré and their Dykeland Legacy*.[9]

Reading *Against the Tides* also reminds me of how much of the contemporary landscape I failed to see and understand in 1969. The copper or cast-iron clapper-gate valves and machined timbers of many an *aboiteau* made it clear that these were not eighteenth-century structures (although their design and function were essentially similar to those of earlier times). By contrast, I took a bus across the Petitcodiac Causeway without realizing that it was barely a year old. As I travelled the Trans-Canada Highway from Amherst to Sackville, I was oblivious that it bridged the Tantramar River atop a dam built in 1960 that had, for ten years before that, been the subject of local resistance because it posed a threat to traditional systems of land management. There were signs, here and there, referring to the MMRA, and the acronym came up (often somewhat dismissively) in conversations with dykeland farmers. But I was not in the Maritime provinces to become embroiled in politics. At the Fort Beauséjour–Fort Cumberland National Historic Site, I figuratively turned my back to the Missaguash River, once the boundary between France and England in this part of the continent, and looked westward, trying to imagine the landscape of the 1750s and 1760s as it would have appeared to French and then English soldiers in the garrison. I blanked the highway and the railroad from my mind's eye but fell far short in attempting to account for the ecological effects of those structures. Fifty years on, Ronald Rudin's book and its thorough explication of the intricacies and consequences of the MMRA's work reminds me of these early inadequacies.

These pages also call forth pleasant recollections. There is a photograph on page 133 in this volume that encapsulates my memories of the marshlands landscape, as I encountered it in 1969 and as I came to know it in different lights through several subsequent visits. The elements of this image – the flat horizon, the expansive sky, the strong and thick growth of grasses, and the dense and distinctive scatter of hay barns once so typical of this and other marshland areas – stand as emblematic.[10] But the landscape was not ever thus. Acadians and their immediate successors took a variety of crops from their dyked fields. Late in the nineteenth century, large herds of beef cattle grazed the marshes, and hay for local consumption was stooked and carried to upland barns.[11] Then western (prairie) producers began to dominate the markets to which Maritime farmers had shipped their meat.[12] The local agricultural economy adapted, turning to supply New Brunswick woodsmen, and Cape Breton, Halifax, Saint John, and other urban centres with vast quantities of hay, before internal combustion engines displaced horses as sources of motive power. Years of economic depression and war further taxed the viability of Maritime farming. Even as Robert Cunningham made his evocative image looking west along High Marsh Road, most of the barns, in which hay had been stored before being loaded onto wagons and railcars for shipment, were empty, rendered obsolete by shifting economic and social forces.[13] Today, relatively few remain. Abandoned structures collapsed, burned, or were stripped to provide the weathered boards formerly much prized as decorative elements in proliferating suburban homes. "*There is / a sense of history here and all / across this marsh*" declared scholar and poet Douglas Lochhead (one of many creative souls who found their muse in the marshes) in the 1970s, as he strode the High Marsh Road thinking of the Acadian who "*sweated and froze in the / ever-wind to make these dykes.*"[14]

Change is the essence of history, and *Against the Tides* is an important contribution to understanding the intriguing, inescapable, ever-changing face of the Maritime marshlands. As I used the photograph on page 133 to mark one facet of the story that this book helps to unfold, readers might turn to others among the book's illustrations for a compelling encapsulation of one of its major themes. Compare the images on pages 7, 44, and 52 with those on pages 49 and 86. The first set shows groups of men with little equipment other than shovels and sledgehammers, some of them knee deep in heavy, cloying marsh mud. Their own muscles, and those of the horses that appear alternately posing with their handlers or straining every sinew as they haul material upslope, are the main sources of power applied to the task of building or repairing dykes and aboiteaux. The situations of men and beasts seem precarious: finding footing on unstable ground; perched perilously on incomplete structures straddling turbulent water; or in danger of pitching, catastrophically, from a steep and narrow slope. For whatever reason, the workers seem engaged, watchful, and in the moment. In the second pair of photographs, machines dominate the mud and the scene; men are scarce and dwarfed by the equipment; their gazes seem to be directed "offstage" rather than to the immediate task and mud at hand, as though they seek or are being given direction.

There is a world of change in these contrasting images – and more in the historical argument that they represent. The first set of images, characteristic of the early twentieth century, bespeaks tradition, local knowledge, and intimate familiarity with one's surroundings. Here men come together to pool their labour, share their knowledge, and work alongside one another for communal ends in ways passed down the generations, from father to son and neighbour to neighbour. The heavy equipment that towers over all else in the second set of photographs is one element in an array of technique – encompassing science, engineering, politics, and finance – brought together to constitute a powerful system of resource and environmental management that betokens the thoroughgoing industrialization of nature. Paul Josephson, a historian of science and technology, and student of Russian and Soviet history, has called such combinations, "brute force technologies."[15]

In a similar but more widely known argument, the anthropologist and political scientist James C. Scott refers to this amalgamation as *high modernism* – "a muscle-bound version of self-confidence about scientific and technical progress" – and insists that it characteristically devalues or discounts the "skills, knowledge, and insight" of local people in its drive to rationalize and centralize in quest of economic efficiency.[16] Yet Rudin refuses, with others, to take this formulation uncritically. His book revealingly complicates the relationship between local and expert forms of knowledge by anchoring both in place. Yes, some traditional ways of managing the marsh were lost, some dykeland farmers felt themselves sidelined as their interests were marginalized, and some MMRA schemes were environmentally detrimental. But traditional practices had evolved over the decades, the second quarter of the twentieth century had been hard on many marshland farmers, and some of the new ways were beneficial.

Context, an artifact of place and time, was and is (as Rudin reveals) vital to understanding the MMRA and its legacy in Maritime Canada.

The importance of setting and circumstance to the story told here is clear from the first pages of Rudin's discussion, in his treatment of the film *Les aboiteaux*. Released in 1955 to coincide with the bicentenary of the Deportation, this twenty-two minute docudrama – substantially the creation of a pair of ambitious young Acadians, the filmmaker and poet Léonard Forest, and his researcher (and future Governor-General of Canada) Roméo LeBlanc – was both a product of its times and a statement for the ages.[17] Since the nineteenth century, a quiet consensus had made space (in New Brunswick in particular), for a small Acadian elite on the fringes of Anglo society and its institutions, even as most Acadians were consigned to traditional, marginal ways of living. After the Second World War, however, this *bonne entente* began to chafe, as social and economic changes remade the countryside, and the need for new accommodations and a new social order gradually came to be understood.[18] In this context, *Les aboiteaux* built a bridge between tradition and modernity. Both were honoured. Tradition had its value. When a storm threatened to breach the dykes, devout men left their community worship to toil together in the foulest of weather in an effort to save their collective patrimony. They did so at the summons of a traditional office holder, chosen "to watch over the aboiteaux" ... "to watch over the dykes." But the emergency was not solely attributable to the storm. The old ways were not what they once were. "They wouldn't listen when I told them that the time had come to make repairs," intoned the lord of the marshes as he made his rounds, "it will soon be too late."[19]

Young men had left family farms for jobs elsewhere, farmers took off-farm work to make ends meet; many of those who remained in the struggling rural economy were old and worn down. Traditional ways were crumbling. As the despairing men face defeat before the rising waters, the traditional overseer of the marshes summons aid in the form of a powerful bulldozer dispatched by the recently arrived MMRA. The dykes are bolstered, the day is saved. But the old structures remain vulnerable. New dykes and aboiteaux, built in new ways with new tools, must be erected. The thoroughly modern MMRA has the mechanical means to do this. But modern knowledge and machinery is, in the end, inadequate for the task of taming the particular tides of the Bay of Fundy. The local knowledge and accumulated wisdom of the aging Acadian overseer proves essential to the proper and secure placement of the new aboiteau. This is a story of past and present brought together to limn a future in which tradition and change are reconciled. Applying the words of an early-twentieth-century ditty to the balance sheet of profit and loss surrounding the MMRA and its legacy, one might well incline to perch on the fence (or dyke) and reflect: "*sooth of lose-and-win; / For 'up an' down an' round ... goes all appointed things, / An' losses on the roundabouts means profits on the swings!*"[20]

Against the Tides is divided into two major parts, labelled "Second Nature" and "Third Nature"; "First Nature" also appears in these pages, but as a necessary foil to the others. Rudin uses these terms heuristically and helpfully. But perhaps a further word or two might assist readers mindful that Henry David Thoreau's celebrated mid-nineteenth-century "word for Nature" was a paean to "absolute freedom and wildness" and that literary theorist Raymond Williams ranked "Nature" as one of the most complex words in the English language.[21] Notions of "First" and "Second" Nature are more commonly invoked than are notions of "Third" Nature. In broad terms, first nature was what Thoreau had in mind: wilderness, pristine territory unmarked by humankind. The realm, some would say, of the gods. Second nature is first nature transformed by human hands. The Roman statesman Cicero gave the idea timeless expression when he wrote in *De natura deorum:* "We sow corn, we plant trees, we fertilize the soil by irrigation, we dam the rivers and direct them where we want. In short, by means of our hands we try to create as it were a second nature within the natural world."[22] In geographer Carl Sauer's classic formulation of 1925, the natural landscape (what some would call first nature) is an assemblage of rocks, soils, landforms, vegetation, weather, and climate developed by the operation of physical processes. The complex of features (such as farms, fields, houses, roads, monuments, and so on) created from, and in the natural landscape over time by people acting according to their needs, beliefs, and means (i.e., their culture) constitute what has come to be called the cultural landscape, or second nature.[23] Sometimes used, haltingly, by writers of the European Renaissance, to characterize the formal gardens then in vogue, the term "third nature" has had limited recent purchase beyond literatures on garden design and evolutionary studies of human behaviour (which class first nature as matter produced by the Big Bang, second nature as biological life, and third nature as ideology and cultural artifacts).[24]

Rudin is more specific than most in using each of these terms (and certainly steers well clear of equating them with matter, life, and culture). His first nature is found where tides move freely across the marshes; his second nature lies behind – is created by – dykes and aboiteaux; and his third nature results from the dams built across a handful of tidal rivers during the latter stages of the MMRA's mandate. This tripartite division adds intelligibility to his argument. In this telling, the salt marshes are pristine, the dykelands are local manifestations of Cicero's fields (without trees), and river dams and the ecological transformations they produce stand equivalent to the perimeter walls and tightly controlled spaces of Italian formal gardens. With these distinctions in mind, the logic of the narrative offered in these pages is clear. Over centuries, even millennia, human activities on the marshlands have been driven by different types of knowledge. Changes in marshland ecologies have been measured by the benefits humans believed they would derive from them. But peoples' "value-judgments" differed and communities came to hold different views of the worth of particular arrangements over time. So ecosystems and landscapes changed.

In the face of such complexities, periodization and categorization help to clarify understanding. But classification is notoriously difficult, especially as it generally reflects the purposes for which it is undertaken. Categories leak, different observers have different perspectives, and questions of scale and abstraction compound the challenges of agreement. Landscapes evolve constantly, and are almost infinitely varied. Dividing lines are hard to fix on the continuum between pristine and highly engineered landscapes. In declaring *The End of Nature* in 1989, Bill McKibben was signalling that human impacts "on the environment, which before had ended at the edge of our villages or the margin of our fields, was now ubiquitous" – and they had been trending that way long before accelerating climate change spurred McKibben's pen.[25] Conversely, wild things – animals, birds, plants, insects – live amid the concrete and glass, the noise and pollution and vehicular traffic of our most dense urban centres. These urban dwellers do not live in exactly the same ways as their conventionally "wild" counterparts in field and forest, but the same can be said of city folk and their country cousins. Adaptation bends and blurs some, but not necessarily all, boundaries. The lives of humans and nature are inextricably entangled.[26] Landscapes are hybrid things, differently understood, and Rudin's First, Second, and Third Natures are, inevitably, simplifications made to elucidate his particular argument about the work of the MMRA.

In pondering the legacy of the MMRA, as Rudin does in the latter part of *Against the Tides,* the challenges inherent in all of this become clear. In the last years of their organization's existence, MMRA leaders pursued the construction of dams across some of the region's rivers with almost monomaniacal zeal. Engineers ignored the concerns of fishery scientists and of local landowners in their drive to hold back the tides and avoid the ongoing demands of dyke and aboiteau maintenance. The ecological consequences were significant. The age-old practice of allowing occasional tidal incursions to fertilize dykeland was foreclosed; fish stocks were depleted; sediment deposition changed the hydrography of tidal reaches below the dams and headponds expanded above them. But people valued these changes in different ways. Some appreciated – and profited from – the lake that grew above the Petitcodiac Causeway and bitterly resented those who argued for its removal and emphasized its artificiality by calling it "a headpond." But those who wished to restore the endangered river faced their own challenges in establishing a baseline to which they could direct their efforts. Defenders of the river, critical of the "Third Nature" created by the dam, wanted to return the ecosystem to an earlier but not "pristine," state. Their ideal lay in "Second Nature," the recently familiar and considerably "tamed" landscape of dykes and aboiteaux. The colonization, by salt marsh grasses, of new banks of sediment on the lower parts of the Petitcodiac was reviled. But on the Avon, where the rapidly extending acreage of *Spartina* became a "top photo opportunity" and an important feeding ground for migrating birds, it was revered. As much as it might resemble "First Nature" on the fringes of the Fundy marshes before Acadians came ashore, this was also a humanized landscape. Its sediment was laid down because of the dam, and its grasses flourished due to nutrients flowing into the ecosystem from the sewage outflow of a nearby town.

Reflecting on these fascinating aspects of the marshlands of the Canadian Maritime provinces, I am reminded that years ago Stephen Jay Gould identified two ways of thinking about time – as arrow and as cycle. Both metaphors made the long deep past of geological time intelligible, the first by presenting it as a series of distinct and irreversible events, the second by stressing timeless order and law-like structure. Significantly, however, the two "do not blend" though they "dwell together in tension and fruitful interaction."[27] One might identify a form of timeless order in the daily ebb and flood, the eternal cycle, of Fundy tides. But the enumeration – first, second, third – of different landscape assemblages implies the march of time, the succession of discrete periods, or a particular arrangement of things. Can this order be reversed?

Advocates of ecological and landscape restoration seem to suggest as much. But is it possible to rewind the clock, when even remote wilderness reserves bear the human stain in the act (and implications) of their designation? Can we go "back home to the old forms and systems of things which once seemed everlasting but which are changing all the time – back home to the escapes of Time and Memory"?[28] This question seems especially pertinent after reading the final pages of *Against the Tides,* which recount recent efforts to undo some of the work of the MMRA by "re-engineering" landscapes now considered less than ideal – for reasons that have a lot to do with the past.

Douglas Lochhead was right in declaring: "*There is / a sense of history here and all / across this marsh.*" It has deep roots. The famous Canadian author and poet Sir Charles G.D. Roberts, raised into his early teens on the edge of the Tantramar marshes, demonstrated this fact in his 1883 poem "Tantramar Revisited." Juxtaposing images of the landscape stretched out before him with nostalgic evocations of earlier days, he marked the physical act of return to his childhood home by raising questions about the emotional, intellectual, and psychological benefits of that journey.[29] Contemplating "*the present peace of the landscape*" while breathing deep the "*Old-time sweetness*" of the marshes, carried on "*winds freighted with honey and salt!,*" he wondered whether it was better to "*Muse and recall far off,*" to remember rather than see, what "*the hands of chance and change*" had wrought on once-familiar fields. Then, almost a century later, we join Lochhead "*walking the bitten beach*" of the Cumberland Basin and finding "*on looking down, a spade-handle.*"[30] Picking up this "*hand carved artifact*" left by "*some bending toiler*" he feels "*the hand / which made and used it / to make a walkway of dyke*" and is drawn by "*this force of the past*" into a deeply personal reverie: "*I let myself in / and talked to him for days.*"

A keen observer of nature, Lochhead is deservedly recognized as one of the most probing and perceptive of the marshlands' many mappers. Stand with him on the edge of the deep, tidal Tantramar River on a March morning, watching as "*The sea-force moves and makes a landscape / empty in a winter season of searching birds.*" This he says "*is a happening of nature we try to chart / and plot our way to nearer understanding.*" But it is also "*the going on of the place / the floating-by of life, so close, so close.*"[31] Lochhead's poems exude a sense of wonder – he marvels at the sweep of the landscape, at its depth (in both human and geomorphological terms), and at the details of its making: "*There in the banked and thrown ice / the leaking limits of the marsh / move free in their melting, leaving crusts / to the shudder of tidal might.*" The opening lines of "Poet Talking" confess: "*I want first to hear / something the sea says, / something the wind knocks; / touch, smell something / the moment has.*" Only then "*when all is there / then I will take it / tell it again my way.*"[32] And so he does. Replete as they may be with introspective ruminations on the topology and processes of landscape formation, Lochhead's poems also lay bare the close imbrication of humans and nature on the marshes.[33]

For Lochhead, the Tantramar was always more than the flat and "*brooding marsh*" that he walked and loved. Here he recognized also "*the hard edge of man's plotting,*" the statements "*of bridge and rail,*" and the "*lines of poles and staggered fences*" that criss-crossed and patterned the marshes – and gave them deeper significance.[34] His marshes provoked new ways of thinking about land and life in all of their dimensions. Among the dominant features of the marshland landscape as Lochhead knew it were the high-power shortwave radio relay transmitters east of Sackville built during the Second World War and demolished in 2014. From this array of uncompromisingly vertical towers on the horizontal marsh, CBC sent its programming to northern Quebec and Radio Canada International, as well as several foreign shortwave broadcasters, which relayed their transmissions around the globe. Many considered the site an eyesore. But for Lochhead it opened a world of meaning:

> *Tracks*
> *from the Transcanada the tracks of snow and hay*
> *point as fingers to the tailored towers*
>
> *of the CBC. This spring is one of messages*
> *and while the snow's crazy abstract grows*
>
> *a Polish commentator in Montreal this day*
> *sends his toiling words to Warsaw and Gdansk,*
>
> *an Italian sings and through the splits of towers*
> *there are sounds in Amalfi and Palermo*
>
> *so the many messages of men give out*
> *their vibes to make a homesick noise*
>
> *around the world from this brooding marsh*
> *where mole and mouse take cover in their place.*[35]

From the intensely local patterns traced by snow and hay, the poem carries the reader on a journey around the globe and back to the veiled realms of mouse and mole, inviting contemplation of the "mysteries, riches, and wonder" of the universe.[36]

Given the differences between Roberts's and Lochhead's perspectives on the Tantramar – Roberts's contemplative and remote, Lochhead's immersive and embodied – one might wonder whether there is, in truth, a singular "sense of history here."[37] If everyone experiences the marsh in their own way, any sense of its past must be diverse, difficult to pin down and as fleeting as the flame-like phosphorescence or atmospheric ghost light – the will-o-the-wisp – reported, in decades and centuries past, by travellers crossing bogs and swamps at night.[38] If "human experience takes on different qualities of meaning and feeling" then "different places take on different identities for different people."[39] Yet Lochhead's poems (and, it must be noted, the works of many of his contemporary creators of marshland imagery) have helped, through the last several decades, to shape a strong sense of the marsh- and dykelands as singular landscapes.[40] Their combination of ecological sensitivity with the quiet conviction that "we inhabit ... [the] ordinary world as it inhabits us," has resonated with a "deep human need ... for associations with significant places" to foster a widespread twenty-first-century awareness of these locales as places worthy of contemplation, emotional engagement, and protection.[41] All of this has greatly complicated discussions of, and decision-making about these landscapes as they are buffeted by "chance and change" – the biogeochemical, demographic, technological, economic, political, social, and cultural shifts that are forever reshaping the face of the earth.

Today, the spectre of climate change brings the future of the long-occupied, thoroughly remade, and intensely storied Maritime marshlands into sharp focus. "Rising Seas Could Ruin Land Acadians Turned from Marshes to Farms" ran the headline above a *Globe and Mail* story in April 2018.[42] Six months earlier, municipal leaders from Amherst and Sackville had declared "N.S. at risk of becoming an island due to flooding" – noting that a recent study by Ottawa's Working Group on Adaptation and Climate Resilience described electricity transmission systems as well as trade flows through the Chignecto Isthmus (valued at $20 billion a year) as highly vulnerable to sea-level rise and storm surges.[43] Most dykes up and down the Bay of Fundy would prove unfit for purpose – unable to stand against the tides – in the face of a once-in-ten-year storm, reported Danika van Proosdij, a specialist in biophysical processes and environmental analysis from St Mary's University in Halifax. The marsh- and dykelands are clearly cultural landscapes under threat.[44]

The pressing question is how to respond. Recent inquiries note markedly different levels of public understanding of the threat that climate change poses to these landscapes and widely divergent opinions about how to answer it.[45] In general, Nova Scotians seemed "more supportive of improving dykes than of ... coastal wetland restoration" in response to rising sea levels, but their choices were conditioned by a range of considerations from fondness for the present scene through economic, cultural, and recreational concerns, to questions of cost. By the reckoning of one study, the "symbolism and amenity of dykelands" as well as "cultural values and status quo bias," are "clearly barriers to adaptation planning, even when discussing the removal of man-made structures." People, these authors conclude, "have very different and perhaps equally valid ideas about what constitutes a sustainable landscape."

NEITHER THIS ESSAY nor the book that it introduces can resolve this dilemma. But *Against the Tides* carries us a good deal closer to developing a thoughtful response to it. As students of cultural landscapes and heritage preservation recognize and respond to the (potential) impacts of environmental change, they are increasingly drawn to emphasize the importance of adaptation over preservation in their work.[46] Adopting more contextualized, place-based approaches, recognizing that landscapes, the dynamic products of "countless dialogues," are much more difficult to protect than buildings, and accepting that their efforts must incorporate humanistic as well as scientific perspectives, they have sought to transcend an older view of the landscape as a "jumble of objects whose origin, function and relationship to each other ... [were] mysteries."[47] Landscape "narratives" have come to be seen as a powerful tool for understanding the dynamics of landscape change and developing "opportunities for creative future transformation of landscapes based upon an understanding of their transformations in the past."[48]

By detailing the work and reflecting on the legacy of the MMRA, *Against the Tides* fills a significant gap in our understanding of the evolution of the Fundy marshlands. It is the keystone in an arch of scientific, historical, and creative scholarship *"taking the measure of nature and man"* that recounts *"not an emptiness, but a loud place / for reading, for noting this and that"* and reflecting on the development of these places from early post-glacial times to the present.[49] Taken as a whole, this story is one of multiple intertwining systems and perspectives: geological, botanical, zoological; spatial, economic and political; visual, symbolic and affective; of tides and traditions; of poets and painters among engineers and planners.[50] In the end and above all, however, it is an environmental story of people and place, of war and peace, of struggle and resilience, and of science and humanity. With Rudin's account filling out our understanding of the most recent decades in this long saga, we are encouraged and enabled to reflect on the sweep of time and tide across the Fundy marshlands; to contemplate anew the variety, intricacy, and consequences of human-environment relations; and to ponder – with some urgency – the future of our planet.

PREFACE

In Search of the MMRA

THIS BOOK FOCUSES on the impact of the Maritime Marshland Rehabilitation Administration (MMRA), which left an indelible mark across roughly 80,000 acres at the head of the Bay of Fundy. When these lands, in present-day New Brunswick and Nova Scotia, were settled by Acadians in the seventeenth century, they included large stretches of salt marsh, created by the largest tides in the world that, twice daily come up the bay, and the rivers that flow into it. These tides deposited large quantities of silt that allowed the creation of marshes that formed a rich ecosystem capable of supporting a wide variety of living creatures. Indeed, the Indigenous people of this region, the Mi'kmaq First Nation, built their lives around these marshes, harvesting the waterfowl and animal life that thrived there; and the Acadians briefly did the same before deciding that they would be better served by draining the marshes, turning them into arable fields.

The Acadians worked collectively to build a system of dykes that held back the tides, and *aboiteaux,* sluices built into the dykes that allowed fresh water to drain from the marsh at low tide, while preventing salt water from entering at high tide. The farmland that was created, commonly referred to as dykeland, still exists today in parts of the two provinces, but the Acadians no longer own more than a very small part of it, most of them having been deported in the mid-eighteenth century. Their lands and the practices that they followed were taken up by waves of English-speakers, who created local organizations, known as marsh bodies, to manage the protective structures. In the decades after the First World War, however, economic circumstances made it difficult for farmers to keep up the dykes and aboiteaux, and by the late 1930s and early 1940s, reports were being sent to all levels of government about farmland that had washed away, or had gone "out to sea," to use the expression of the time.

In response, the Canadian government, in collaboration with the two provinces, created the MMRA in 1948 to shore up the existing structures, replacing or repairing 373 kilometres of dykes and over 400 aboiteaux. Before winding up its operations in 1970, the MMRA had also constructed five tidal dams, blocking a number of the major rivers that flowed into the Bay of Fundy so that the tides could no longer go upstream, in the process precluding the need to maintain the dykes. Before the MMRA was done, its projects had impacted nearly 500,000 acres of farmland, both the drained marshes and the upland sections of farms that depended on the marsh so that the farms were viable, and had influenced the lives of roughly 200,000 people, a number that spiked as a result of the construction of a tidal dam at Moncton, one of Atlantic Canada's most important cities.[1] Most of these individuals had nothing to do with agriculture, but their lives were affected by the MMRA just the same.

In spite of the MMRA's role as a major postwar initiative, little has been written about the agency since it closed down over fifty years ago. The agency has remained in the shadows even though there is a significant international literature dealing broadly with wetlands, environments where land and water meet. While wetlands can take on a variety of forms, from salt marsh (as in the Bay of Fundy region) to swamps and bogs, all of these environments have been in decline for centuries, one researcher estimating that nearly 90 percent of the world's wetlands at the start of the eighteenth century were destroyed in the three centuries that followed; and a similar decline has been calculated for the salt marshes around the Bay of Fundy since the arrival of the Acadians. This precarious situation received international recognition in 1971 with the drafting of the Ramsar Convention on Wetlands, whose signatory countries agreed to promote the conservation of these landscapes.[2]

In the context of the challenges faced by wetlands, a substantial literature has emerged. Some authors, for instance, have pointed to the value of the food produced in such environments, explaining how wetlands "produce a total biomass close to that of tropical forests and far in excess of agricultural fields." Still others have focused on the various threats to the survival of wetlands, identifying economic pressures as well as the cultural prejudices that have labelled them as "unvaried stagnant places" that were "denigrated as uninteresting, unhealthy environments."³ In addition, there is a body of literature that runs against a declensionist narrative, pointing to the opportunities for restoring wetlands that have been damaged or destroyed and protecting those that have remained intact.⁴

Those who have written about the marshlands of the Bay of Fundy region have picked up on these larger trends in the wetlands literature, tracing how the marshes were drained, how the dykes and aboiteaux deteriorated in the early twentieth century, and how those protective structures were reinforced in the years after the Second World War. Many of these works are discussed in the pages that follow, but a significant number make no reference to the MMRA, even though it transformed both how the dykelands were protected and the role played by farmers in the process.⁵ In still other cases, authors have referred to this agency, but only in passing. For instance, Harry Thurston, who has written extensively about environmental issues in the Bay of Fundy region, has pointed to the activities of the MMRA, questioning many of its initiatives because "fighting the tides inevitably set into motion a sequence of unforeseen, and often intractable, ecological consequences."⁶ Taking a different tack, the government-sponsored *Maritime Dykelands: The 350 Year Struggle*, provides, as the title suggests, a particularly heroic account of efforts "to wrest land from the sea."⁷ As for the MMRA, by the time it had finished its work, "the twenty year war to save the region's most fertile soil had been won."⁸

Various perspectives on the legacy of the MMRA can be drawn from the studies devoted to the marshlands of Nova Scotia and New Brunswick, but what all of these works have in common is their limited account of the role of the MMRA in shaping both the region's landscape and the lives of its people. To a considerable degree, this situation can be explained by the fact that the paper trail chronicling its history seemed to have disappeared when the MMRA shut its doors. When I began this project, I figured that a federal government agency that operated for over two decades would have left an institutional record with Library and Archives Canada (LAC), but while there is a significant body of material at LAC dealing with the process leading up to the MMRA's creation, the trail goes dry once the agency came into existence.

After running up against numerous dead ends, I learned that at some point after 1970, when responsibility for the structures repaired or constructed by the MMRA was handed over to the provinces, two trucks pulled up to the agency's head office in Amherst, Nova Scotia, in the middle of marsh country and effectively on the border between the two provinces. One truck took the Nova Scotia–related documentation to Truro, long the centre for agricultural affairs in the province and home to both departmental offices and the agricultural college. Here I found the MMRA files in a basement location that, as far as I could tell, had been untouched for decades. The other truck went to Moncton, where the New Brunswick documents were deposited. I found them in a shed behind the provincial agriculture offices, which were accessible only when the snow that surrounded it for six months of the year had melted. Inside the shed, the documents were in boxes on shelves that could be reached only by climbing over picnic tables and barbecues.[9]

However challenging this adventure might have been, I was able to access the provincial collections only with the support of Claude Robichaud in New Brunswick and Kevin Bekkers in Nova Scotia, who are responsible for the upkeep of the protective structures, and are ever mindful of the combined impact of climate change and the sinking of the region's land mass that requires that dykes be built higher than had been the case in the past. I was also assisted by Melanie Lucas, a records analyst with the Nova Scotia Department of Agriculture, who was responsible for removing the documents from their basement location and cataloguing some of them for consultation at the department's offices in Truro. Still other Nova Scotia material, which never made it to Truro, remains in the possession of the province's four aboiteau superintendents, who provided both access to the documents and much helpful advice on matters relating to dykeland management. Many thanks to Dave Smith (in Kentville), Darrell Hingley (Truro), Craig Bauchman (Windsor), and Gary Gilbert (Nappan).

In addition, I was helped, in more conventional but still important ways, at various archives across Atlantic Canada. Special thanks go out to Rob Gilmore at the Public Archives of New Brunswick, and to archivists at the two marshland universities: David Mawhinney at the Mount Allison University Archives, and Catherine Fancy and Pat Townshend at the Acadia University Archives; and at the Kings County Museum, I was helped out by its curator, Bria Stokesbury. I also received significant support from Leah Rae, an archivist for Library and Archives Canada based in Dartmouth, Nova Scotia, who helped me secure access to regional material pertinent to the connection between the MMRA's tidal dam projects and the federal Department of Fisheries and Oceans (DFO). Further DFO documents were secured by way of an Access to Information request for material pertaining to the Petitcodiac Causeway.[10] Still other requests with regard to the Petitcodiac project yielded documents from the New Brunswick government and from the municipalities of Moncton and Riverview.

The archival record extended beyond government offices and formally constituted archives, as I also unearthed documents in the homes of marsh owners in Nova Scotia, who were in possession of the minute books of marsh bodies, the local organizations that were responsible for maintaining the dykes and aboiteaux prior to creation of the MMRA.[11] In a few cases, these records stretch back to the nineteenth century and continue to the present. By and large, however, the minute books become less detailed following the start of the agency's operations as control passed from the community to MMRA headquarters in Amherst. In fact, in some cases, where dams were constructed, the marsh body minutes came to an abrupt close, as the proprietors had no further role to play once the tides could no longer reach their properties. The question of the decline of local control with the rise of the MMRA is a central theme in this book, but for now, I want to thank the following individuals (with the name of their marsh body in parentheses) who opened their doors and frequently fed a solitary researcher on the road: Andi Rierden (Dentiballis); Jim Inglis (Le Farm); Larry Hudson and Peter de Nuke (Queen Ann); Eric Patterson (Bishop Beckwith); Philip Davison (Falmouth Great Dyke); Tara Hill (Victoria Diamond Jubilee); Charlie Curry (Horton).

In the end, however, the documentary record could take me only so far, and so I also benefited from the collaboration of individuals whose lives had been impacted in one way or another by the MMRA. They took me out of the archives, on to the land and sometimes into the mud, because this is fundamentally a story about individuals living within a particular environment. As is standard operating procedure for oral history projects, I think of these people as collaborators who worked with me to create testimony that had not previously existed. Some of them explained to me the intricacies of dyke and aboiteau construction, not to mention the upkeep of ditches that allowed water to drain from their lands; still others showed me the impact (both positive and negative) on the environment of the MMRA's tidal dam projects, which stretched far beyond drained farmland. And, once again, a number of these individuals fed me. So heartfelt thanks to: Anna Allen, Roger Bacon, Jim Bremner, Alyre Chiasson, Graham Daborn, Michel DesNeiges, George Dorsay, Shannon Douthwright, Robert Ettinger, Gary Griffin, Nancy Hoar, Hank Kolstee, Mary Laltoo, Daniel LeBlanc, Victor LeBlanc, Wiebe Leenstra, Robert Palmeter, Ernie Partridge, Arthur Phinney, James Reicker, Ann Rogers, Jim Sellars, Al Smith, Harry Thurston, George Trueman, Danika van Proosdij, John Waugh, Bob Wilson, John Wilson, Jim Wood, and Sonja Woods.

These individuals' testimony plays a central role in the pages that follow, providing a human connection to this story, which, from the archival record, can sometimes be reduced to engineering reports; and many of them participated in filming the short documentary *Unnatural Landscapes*, which I produced and which was directed by Montreal filmmaker Bernar Hébert. The film, which complements the book, is available at http://unnaturallandscapes.ca, a website created by Antonia Hernández and Corina MacDonald. Both the website and the book have also benefited from original sketches created by Montreal artist Caroline Boileau. I have now had the good fortune to work with Bernar, Antonia, Corina, and Caroline on a number of projects, and have always been impressed by their creativity, turning my half-baked ideas into works of art.

The list of people who befriended me along this journey through a special landscape is long, but one person deserves special thanks for his willingness to share with me his unique knowledge about the marshlands, the dykes that protected them, and the role of the MMRA in the region. Since this book is to a large degree about different forms of knowledge, I was fortunate to have been led (by many people) to Ernie Partridge, who was one of the agency's first employees, working for it from 1949 to 1962.

Born in 1929, Ernie grew up in Dorchester, New Brunswick, and became fascinated as a boy by the coming and going of the tides that he could see along the Memramcook River, which flows behind his property. Still in his late teens, he came to his job with an understanding of how dykes had been built for centuries, and was able to claim to me, while brandishing one of his prized dyke spades, that he was the only living person who had such knowledge. He worked his way up the ranks within the MMRA, ultimately becoming responsible for the upkeep of the dykes and aboiteaux from the Nova Scotia border to Moncton, a position that had him walking over 160 kilometres per week to check on the condition of these structures. By the early 1960s, having done his job so well, there were no further dykes to fix and he moved on to work for the federal penitentiary that was only a kilometre from his home. Over fifty years later, he still describes his time with the MMRA as the best job of his life, an enthusiasm that he communicated to me in six separate interviews. In my practice as an oral historian, I have had the good fortune of meeting a number of born storytellers who were nothing short of inspirational. Ernie Partridge, to whom this book is dedicated, is at the head of that short list. His muddy fingerprints are all over this book.

Of course, I wouldn't have been able to visit any archival holdings, interview any of my collaborators, or produce *Unnatural Landscapes* without the financial support of the Social Sciences and Humanities Research Council of Canada, which has generously funded my various projects over the past forty years. I also benefited from the constant support of my colleagues at Concordia University's Centre for Oral History and Digital Storytelling, whose activities have inspired me to think about both the creation of testimony and its presentation to the public.

In addition, I received encouragement from early in this project from UBC Press, and more specifically from James McNevin, who helped me steer a manuscript through the long road to publication during a pandemic. Indeed, this will always be my pandemic book, which I submitted to James in its original form on the morning of March 13, 2020. Later that same day, everything in Quebec was locked down. In that context, I am very grateful to the two anonymous readers, who somehow found the means of staying sufficiently focused to offer helpful suggestions during the days of confinement.

Last but not least, my thanks to the other writers in the Rudin family: my wife, Phyllis (the novelist), and son, David (the journalist). It was always comforting to have company when writing problems emerged or the inevitable writer's block struck.

In the end, of course, I am solely responsible for what follows. Nevertheless, books like this one are collaborative efforts, even though only one name is on the cover. As we will see in the pages that follow, marsh owners similarly needed to work together to protect their lands from the tides.

A Note on Sources

THIS BOOK LEANS heavily on documents that ended up in the possession of the two provincial departments of agriculture. However, very few of those documents have been catalogued, making it difficult for me to provide references that would enable subsequent researchers to follow my lead. As a result, in the references to those unarchived documents, I have simply referred to them as being held by either the Nova Scotia Department of Agriculture (NSDA) or its New Brunswick counterpart (NBDA). In the former case, while most of the documents can be found in Truro, I have indicated a more specific location when held in offices in Kentville, Windsor, or Nappan. All of the New Brunswick documents are available at the Department of Agriculture facilities in Moncton.

The pages that follow also make frequent use of the minute books of the Maritime Marshland Rehabilitation Administration (MMRA) Advisory Committee, referenced in the notes as the MMRA AC (followed by the date of the meeting). Researchers who want to consult these documents can find them in two locations. A complete set of these minutes can be found at Truro, included among documents that have been entered into a database by the Nova Scotia Department of Agriculture. In addition, a nearly complete set is available at Library and Archives Canada.[1]

While most of the sources used in preparing this book were in English, some were in French, as befits a landscape that was created by settlers from France and that has had meaning for Acadians. In those cases, I have translated the sources into English to make them as accessible as possible.

Against the Tides

PROLOGUE

Whose Knowledge? Which Nature?

ONCE AN ACADIAN LANDSCAPE

ONE MORNING IN August 1954, Ernie Partridge, the chief of the Dorchester, New Brunswick, volunteer fire department, assembled his crew to take part in a film shoot in the nearby southeastern New Brunswick town of Memramcook. *Les aboiteaux,* created by pioneer Acadian filmmaker Léonard Forest, was the first film ever made with an original Acadian screenplay; the story it told was a profoundly Acadian one, dealing with the physical object at the heart of Acadian settlement along rivers that flowed into the Bay of Fundy, in present-day New Brunswick and Nova Scotia.[1]

Unlike other European settlers, the French colonists who came to this part of the world in the seventeenth century only rarely felled trees to secure access to arable land. Rather, the Acadians, drawing on the knowledge they brought with them from Europe, set out to drain the salt marshes that were flooded twice daily by the tides of the Bay of Fundy, the largest in the world. They constructed a system of dykes and drainage channels (aboiteaux) that were capped with a hinged gate (*le clapet*) that allowed fresh water to flow out at low tide and prevented salt water from flowing in at high tide. Within a few years, the marshes were drained, transforming the rich ecosystem of marsh grasses that had supported life, both human and non-human, for centuries. In its place, there was now arable land, protected from the tides. For their efforts, the Acadians became known as *les défricheurs d'eau,* clearers of water.[2]

But much had changed by the mid-1950s. The Acadians were deported in the mid-eighteenth century, so that only a relatively small part of the roughly 80,000 acres of drained marshland still under cultivation was owned by French-speaking farmers, mostly in the vicinity of Memramcook, the site of the film shoot, which emerged as an important centre for the *nouvelle Acadie* that emerged after the deportation, serving as home to the first Acadian institution of higher education and earning the title of *le berceau de l'Acadie* (the cradle of Acadie). For nearly two centuries, marshland across the two provinces, as depicted in Map 1, had largely been owned by English-speakers who adopted the Acadians' practices to keep the protective structures, the dykes and drainage channels, in good repair. They even referred to these structures as aboiteaux (usually pronounced *abideaux*), thus making the Acadian term their own.[3]

By the 1920s, however, these devices had begun to decay due to the decline in farmers' incomes, a situation linked to the reduced demand for hay, one of the major crops grown on this land, following the rise of the internal combustion engine; and the ability of farmers to look after their lands was further compromised by the Great Depression. By the late 1930s and early 1940s, farmers were writing to government officials about their desperate situation, describing land that was literally washing away. One such message described how "a large acreage of our dykelands in the Tantramar District [of New Brunswick], are in an unproductive state, and that under present marketing conditions farmers are unable to bear the necessary expense of draining and dyking." The farmers hoped that Ottawa would "cooperate with us in improving the conditions of said lands."[4]

The farmers' lobbying ultimately bore fruit in 1948, when the federal government created the Maritime Marshland Rehabilitation Administration (MMRA), whose activities are the focus of this book. This is the first extended analysis of the legacy of the MMRA, which ceased to operate in 1970 but whose impact on the environmental and cultural life of the region lives on to this day. Marsh owners may have simply sought support for dyke repairs, but instead secured an agency that went much further, constructing tidal dams across the region and more generally bringing to an end the three-centuries-long tradition of local management of the marshes.

The MMRA's history highlights the complicated relationship between local and expert forms of knowledge. This relationship is at the heart of this book. Across the region, marsh owners were marginalized by the agency's

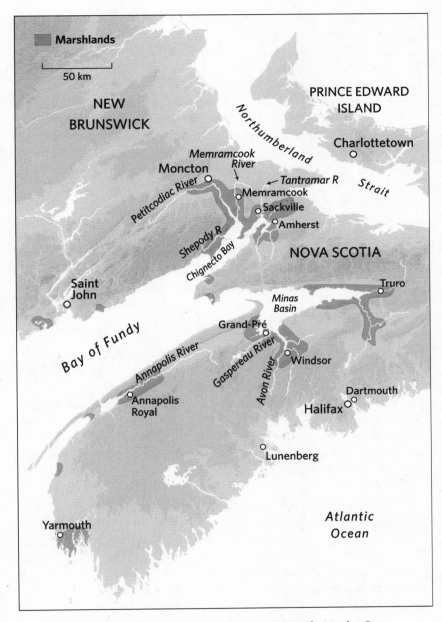

MAP 1 Marshlands in Nova Scotia and New Brunswick. *Map by Marilyn Croot; adapted by Eric Leinberger.*

engineers, but this did not mean that their knowledge was always pushed to the side. In the case of the tidal dams, such local knowledge was frequently ignored, resulting in significant environmental damage, but in terms of the reconstruction of the dykes and aboiteaux, their know-how continued to be valued, as we see in *Les aboiteaux*.

The fictional film deals with an Acadian community that was no longer able to protect its dykeland from the tides. It has a certain documentary feel, as all of the roles are played by individuals from the community, over a hundred in total. We see young men boarding buses to find work elsewhere, their fathers left behind to cope as best they can; and we see Placide Landry (played by retired butcher Adolphe E. Leblanc), the individual responsible for the structures (known in French as the *sourd des marais* [lord of the marshes]), trying to cajole them to do their part, but with little success. We are left with the impression, often communicated by the leaders of the MMRA, that more conscientious stewardship of the aboiteaux might have enabled the community to avoid the crisis to come.

Fearing the worst, Placide looks for ways to bolster the dykes and so travels to the MMRA headquarters in Amherst, Nova Scotia, only thirty-five kilometres from Memramcook and effectively straddling the border between Nova Scotia and New Brunswick. In order to fulfill its mandate to reconstruct the structures holding back the tides, the MMRA designed sluices that had scarcely changed for centuries, now using prefabricated timber for the channel and replacing the wooden gates with bronze so that they would last much longer. In *Les aboiteaux,* we see Placide getting a tour of the MMRA factory, part of a sprawling complex in Amherst, where the agency's professional staff developed new techniques and products to protect the marshlands, and where, during the winter, Ernie Partridge, the head of the Dorchester volunteer fire department, worked in the early 1950s.

Growing up along the Memramcook River, Ernie would regularly pass the time at a canteen that his father operated along the wharf at Dorchester, where the crews on docked steamers would stop to eat. "As a ten-year-old, eleven-year-old kid, when you see the tide in full, that big body of water, I couldn't figure out how that little bit of mud, forming a dyke, would hold back that tide. I was completely amazed with that, ever since I was a kid."[5] This childhood fascination turned into Ernie's dream job when he was hired in 1949 as one of the MMRA's first employees, taking work crews out to repair dykes and aboiteaux that had failed. But during the winter, in his first years with the agency, he worked at the facilities in Amherst, helping the engineers with tests they were carrying out to see how well

Whose Knowledge? Which Nature?

FIGURE 0.1 Trying to hold back the tides. Les aboiteaux, *copyright 1955 National Film Board of Canada. All rights reserved. Collection: Cinémathèque québécoise.*

water passed through new sluices they were designing. If Placide had been a real person, he might have run into Ernie.

In *Les aboiteaux,* we see Placide return from Amherst to try to convince his neighbours to work with the MMRA to rebuild their dykes before disaster struck. He describes to them how the MMRA would provide this work free of charge, which was true, with the condition that they look after the management of the ditches on their lands that would lead water to the aboiteaux. But they refuse to believe Placide, and mock him for being so gullible, in the process adding to the impression that they were at the root of the problem.

Placide's rejection is the set-up for the film's most dramatic moment, when the dykes actually fail in the midst of a torrential storm. The entire community is crowded into the parish church when Placide comes with the bad news. The priest brings Mass to a close and tells his parishioners that they need to try to save the dykes. As we see the marsh owners working hard in the driving rain, action that is depicted in Figure 0.1, Ernie Partridge and his volunteer fire department were off-camera providing the precipitation from their hoses. So here was Ernie, who repaired the dykes for the MMRA in real life, helping create the environment in the film for the dyke's failure.

Recognizing that he and his neighbours were powerless to save the day, Placide telephones his contacts back at the MMRA, who send a bulldozer to move land so that the dykes could survive, at least for the moment. On its arrival, the operator is treated like a hero, in contrast to the marsh owners' mocking dismissal of MMRA assistance when proposed by Placide. This is only a temporary solution, however, and the film ends with the installation of an entirely new dyke and aboiteau, constructed by Ernie Partridge's MMRA crew.

In order to build the new structures, the film shows local men queueing to secure work, but when the elderly Placide steps forward, he is sent away, suggesting that his local knowledge is no longer required in the new MMRA world, ruled by the expert knowledge of engineers. He walks away disconsolately, figuring that his productive life is over, but is soon summoned back by the work crew, who quickly realize that it requires his understanding of the tides to maneuver the new aboiteau into place. The film closes with Placide's knowledge validated, and with the local community, decked out in traditional Acadian attire and dancing to traditional Acadian music, celebrating its escape from disaster.

Les aboiteaux has a particular Acadian context, and its final scenes speak to the challenges for this people, which, in 1955, was still trying to find its footing on the bicentenary of its deportation. Indeed, the commemorative events surrounding the bicentenary, which included a screening of *Les aboiteaux,* reflected the rise of a new generation, including Léonard Forest, that was looking to modern means to allow the survival of a small population.[6] There were roughly 250,000 Acadians in all of Atlantic Canada in the mid-1950s, a people who had held on tenaciously to certain traditional modes of behaviour, often viewing change as a threat, which was not the case at the conclusion of *Les aboiteaux.*[7] Acadians were on the eve of their own Quiet Revolution, paralleling the one in Quebec through an embrace of modernity. This process, which took shape, particularly in New Brunswick, through such developments as the choice of Louis Robichaud as the first elected Acadian premier in 1960 and the creation of an Acadian university (the Université de Moncton) in 1963, was foreshadowed in *Les aboiteaux* through both the emergence of Forest as a cultural leader and the message that Acadian identity did not have to be threatened by innovation.[8]

But this is not a book exclusively or even primarily about Acadians, even though I only came to the topic in the context of two previous book projects dealing with particularly Acadian landscapes. This accidental Acadian trilogy began when I was drawn to the landscape of Acadie's beginnings on Île Ste-Croix, an island on the international border between New Brunswick and Maine, in connection with commemorative events in 2004 marking the 400th anniversary of the first effort by France to create a permanent settlement in the Americas.[9] Acadians had traditionally viewed the start of their modern existence in the aftermath of their deportation, giving pride of place to locations with a connection to the *grand dérangement* such as Grand-Pré in Nova Scotia, immortalized in Henry Wadsworth Longfellow's poem *Evangeline*.[10] But in 2004 some Acadian leaders saw an opportunity to put their people on a timeline similar to that of other North American settler societies. If the Québécois could trace their beginnings to the establishment of Quebec City in 1608, Acadians could do the same by drawing attention to how settlers spent the winter of 1604–05 on Île Ste-Croix.

Subsequently, I explored a landscape that had once been home to more than 1,200 people, mostly Acadians, who were removed from their lands to allow the federal government to create New Brunswick's Kouchibouguac National Park in the late 1960s and early 1970s.[11] Because it had been the site of a twentieth-century Acadian removal, what some have called "*une deuxième déportation*," the Kouchibouguac landscape became further evidence of a legacy of dispossession. But unlike earlier stories, such as *Evangeline*, which focused on the travails of its stoic heroine, who travelled across North America in search of Gabriel, her love, from whom she had been separated by the British, the Kouchibouguac saga spoke to the resistance of former residents who would not accept their dispossession. In particular, it featured one individual, Jackie Vautour, who refused to leave his land up to the time of his death in 2021, nearly fifty years after efforts to forcibly remove him.

With Vautour's death, no one lives either on Île Ste-Croix or in Kouchibouguac National Park, but this has not prevented those sites from looming large in the Acadian imagination. In a similar manner, relatively few Acadians (the farmers from the Memramcook area aside) have tilled the drained marshland of New Brunswick and Nova Scotia since the mid-eighteenth century, but this has not prevented that landscape from serving as a source of inspiration for Acadians who were mindful that *Acadie* had disappeared from the map following British conquest in 1713, decades before the deportation. Here was an environment that the Acadians had created and that still exists, in the process speaking to something that belonged to them, instead of focusing on their legacy of loss.[12] This environment also had a certain metaphorical power, with the dyke providing protection for Acadians from a hostile world, and the aboiteau with its hinged valve controlling access to Acadian society. In this context, the drained marshland and the objects connected to it have occupied an important role in Acadian cultural history over the past seventy years, starting with Forest's film.[13]

In the end, however, unlike either Île Ste-Croix or Kouchibouguac, the drained marshlands have been continuously occupied by settlers since the seventeenth century. While Acadians were involved with the initial draining of marshes across the region, over the past 250 years the cultivation and upkeep of these lands have largely been the work of English-speakers, who – as we will see in Chapter 1 – generally adopted the Acadians' practices, making it appear as if there had been a seamless transfer from one regime to the next, however cataclysmic the deportation may have been for the Acadians. This narrative of continuity was clearly articulated in the UNESCO text paying tribute to the outstanding universal value of the drained Grand-Pré marshland when it became part of a World Heritage Site in 2012. The citation stressed the timelessness of this landscape, pointing to "the permanency of its hydraulic drainage system using dykes and aboiteaux and its agricultural use through a community-based management system established by the Acadians and then taken over by the Planters and their modern successors."[14]

While other sectors of the same World Heritage Site are more distinctively Acadian in their focus on the deportation, using such devices as the monument to the fictitious Evangeline or the reconstructed church where the very real local population was rounded up, such Acadian distinctiveness is difficult to find in terms of managing the marshes. This point is underscored in *Les aboiteaux* in connection with Léonard Forest's research for the role of Placide, the superintendent of the dykes. As Forest put it in his notes, he spent time "with two elderly aboiteaux builders ... They were both old, both were 82. They had each devoted over fifty years of their lives to the construction and repair of the dykes and aboiteaux."[15] As it turned out, one of the superintendents, Fred Palmeter, hailed from a family that we will meet again in these pages, and that came to Grand-Pré immediately after the deportation, while the other, Théotine Landry, was from the Memramcook valley. In the end, Forest built the character of Placide around the Acadian, but otherwise found the two interchangeable in terms of their role in deploying their local knowledge to manage the drained marsh.

More broadly, the situations presented in *Les aboiteaux* could just as easily have taken place in English-speaking communities throughout the region, in the more than 100 marsh bodies, local organizations that were operating in the early 1950s, during the first years of the MMRA. Many of these communities were dealing with decaying dykes, internal conflicts about how to fix them, and the presence of a federal agency that was poised to alter the relationship between marsh owners and the land, bringing in crews such as the one Ernie Partridge headed and more generally introducing its own expert knowledge into communities that valued their autonomy. Forest's film spoke to a certain moment in Acadian history and to a particular Acadian connection with a landscape they created, but the use by MMRA leaders of the language of expert knowledge was ultimately much more significant than whether the marsh owners spoke French or English.

Whose Knowledge?

In various contexts and in a wide array of national settings, the twentieth century witnessed a contest between long-standing traditions of local control and efforts by experts of various types to assert their authority. *Les aboiteaux* provides the viewer with a rather benign version of such an encounter. We see the resistance of local marsh owners when Placide first tries to encourage his neighbours to work with the MMRA, but in the end the community welcomes the arrival of help from the agency, which in turn needs Placide to complete the installation of the new aboiteau.[16] Similarly, when Ernie Partridge went to work for the MMRA shortly after its creation, he was bringing his own knowledge of traditional ways of building dykes and aboiteaux to an agency that was committed to the use of modern engineering practices.

On the face of it, the interests of marsh owners and the new agency were in sync, but it was clear from the start that the MMRA held the proprietors in low esteem, constantly blaming them for the deterioration of the protective structures. In this context, when the agency embarked on a program of dam construction in the 1950s, it was designed in part to remove the proprietors from the picture by preventing the movement of tides up a number of the major rivers that flowed into the Bay of Fundy, essentially ending the need for dykes and aboiteaux upstream from the dams. With regard to the Tantramar River Dam in southeastern New Brunswick, a feasibility study published in 1953 touted the project, which would allow "responsibility of maintenance (to) rest with a central authority ... Experience has proven that when this responsibility is in the hands of individual owners, or small groups dependent on one another, satisfactory results are not always maintained, much to the detriment of the area and the community concerned."[17]

The "small groups" referred to here were the marsh bodies, local organizations that had been responsible for the upkeep of the protective structures since the late eighteenth century, but whose autonomy was significantly reduced with the creation of the MMRA. Indeed, the agency agreed to repair structures only where the marsh bodies had formally agreed to accept the new rules of engagement. While the older landscape largely remained, as we see in *Les aboiteaux,* there was a significant shift in control over it from local marsh owners to the professionals, mostly engineers, who led the new federal agency.

This contest between different understandings of the environment, between local and expert forms of knowledge, has generated a significant literature, to which this book contributes. That literature has been strongly marked by the writings of James C. Scott, in particular his *Seeing Like a State.* Published in 1998, the book still provides inspiration for researchers trying to understand the connection between different forms of knowledge and their impact on the landscape.[18] In Scott's conception, local knowledge, also referred to as "situated knowledge" or *mētis,* "represents a wide array of practical skills and acquired intelligence in responding to a constantly changing natural and human environment."[19] The crucial matter for Scott was the knowledge of the local environment that enabled individuals to responsibly adjust certain practices in the face of the conditions at hand. He presented, for instance, the example of farming, which might have general rules but which "takes place in a unique space (fields, soil, crops, soils) and at a unique time (weather pattern, season, cycle in pest populations) and for unique ends (this family with its needs and tastes)."[20]

Scott's particular concern, however, was to explore what happened when this local knowledge was disregarded by experts who attempted "a mechanical application of generic rules that ignores these particularities," leading to "practical failure, social disillusionment, or most likely both ... While something can indeed be said about forestry, revolution, and rural settings in general, this will take us only so far in understanding *this* forest, *this* revolution, *this* farm."[21] Scott referred to this type of thinking as high modernism, and much of the literature spawned by his writing has focused on the damage that can be done to both the environment and the people who inhabit it when local circumstances are ignored. He explained that high modernism was advanced by experts who had "made their successful institutional way in the world by the systematic denigration of practical knowledge."[22] Equipped with what Scott called "a muscle-bound version of self-confidence about scientific and technical progress," high modernists, in particular planners of various types, felt justified in transforming the environment, "wiping the slate utterly clean and beginning from zero."[23]

In the process, local knowledge might be pushed out of the way, but this could be justified because, according to Scott, "the logic of high modernism implicitly discounts the skills, knowledge, and insight of those whose future is being socially engineered. Virtually nothing is to be learned from those who are, after all, destined to be the pupils, the trainees, in a scientifically designed social order."[24] High modernists sought the transformation of both distinctive landscapes and the people who occupied them. After all, it was under the watch of those with local knowledge that a need for major change had arisen; or, to put the matter more specifically in the context of the marshlands: why bother fixing the protective structures if the marsh owners were going to be allowed to mismanage them as they had in the past?

In this context, there were numerous mid-twentieth-century projects designed to fix or rehabilitate, to employ the commonly used expression of the time, both nature and society, often combining the two as objects that could be repaired through the use of expert knowledge. This broad application of the idea of *rehabilitation,* dealing with both physical and behavioural improvement, was evident in such grand plans for the postwar world as the United Nations Relief and Rehabilitation Administration (UNRRA). Established in 1943, UNRRA was "concerned not only with relief – that is with the provision of material needs – but also with rehabilitation – that is with the amelioration of psychological suffering and dislocation."[25] This all-encompassing conception of what needed to be fixed was also central to Canada's postwar plans that were crafted by the General Advisory Committee on Demobilization and Rehabilitation. Like UNRRA, the committee was dedicated to dealing with "the whole problem" under the watchful eye of "professionals with scientific credibility."[26]

More specifically in terms of the environment, many of the high-modernist rehabilitation projects of the era were involved, as was the MMRA, with the diversion or control of water, in the process dealing with "mobile nature." As Mark Fiege has explained, much like other forms of nature such as soil or organisms, the tides of the Bay of Fundy disregarded human boundaries, in the process creating an "ecological commons, a mobile nature that in moving across boundaries complicated the fundamental order of the (property) grid by joining fragmented parcels into a larger whole."[27] Discussing the creation of an irrigated landscape in the American West, Fiege described how "nature will always draw us out of our individual plots and, whether we choose to recognize it or not, transform us into groups in which we, as individuals, have standing only in relation to the community."[28]

In practice, however, creating such communal action was not always a simple matter. As Shannon Stunden Bower has observed with regard to draining water from sections of southern Manitoba, there was a "mismatch between liberalism (with its emphasis on individualism and private property) and the wet prairie environment (with its water flows that continue in disregard of property boundaries)."[29] This mismatch could create conflict if property owners did not want to work together, or if the state felt the need to intervene, introducing expert knowledge that undermined more local ways of understanding an environment.

Taken to its extreme, the mid-twentieth century saw high-modernist efforts to transform both the movement of water and the people who needed to be displaced in the process. For instance, with regard to the Tennessee Valley Authority (TVA) of the 1930s, Scott has shown how, besides the creation of hydroelectricity, the TVA also sought the resettlement of "a backward population [to] provide them the means of economic and cultural citizenship." The leader of the project, Arthur Morgan, "was a social planner as well as a physical planner ... The objective was not simply a working dam, but a transformed population."[30]

The TVA model was replicated in Canada through such projects as the construction of the St. Lawrence Seaway in the 1950s. Daniel Macfarlane has described the seaway as "the largest rehabilitation project in Canadian history ... By reorganizing spatial and physical environments, planners aimed to improve the lives of area residents, as well as the residents themselves."[31] Similarly, in the 1960s New Brunswick's Mactaquac Dam project was designed, according to James Kenny and Andrew Secord, so provincial officials might "engage in social engineering as they planned the relocation, retraining, and rehabilitation of people displaced by the hydro project."[32]

The history of the MMRA was marked by a similar dynamic toward the end of its operations when it flexed its muscles and ignored local knowledge to construct tidal dams across the Petitcodiac and Avon Rivers. However, such projects did not *have to be* conceived to pit expert and local knowledge against each other, which is in a sense what we see in *Les aboiteaux* as the MMRA recognizes that it needs Placide's mētis to reconstruct the dyke and aboiteau. Indeed, several studies have pointed to more collaboration than conflict between possessors of local and expert knowledge with regard to "mobile nature." For instance, in the Manitoba context, Stunden Bower described a relatively consensual process in which government experts played a key role in creating drainage districts to look after the province's ecological commons, but only where "there was not substantial opposition mounted by area residents."³³

Donald Pisani has described a similar process with regard to the role played by the US Bureau of Reclamation in reshaping water use, particularly in the American West. Pisani did not see any top-down imposition "of state and federal control over natural resources [that] resulted in a loss of local control." Instead, he argued that "the untold story of the twentieth century is the proliferation of special districts and their relationship to county, state, and federal institutions of government." He described how after the Second World War the number of "special districts more than tripled ... and many pertained to natural resources, particularly flood control and water districts." In the end, the impact of these local institutions, together with American reverence for the family farm, "clashed with the idea of increasing state control over natural resources," tempering its impact.³⁴

Pisani viewed the coexistence of local and expert control as proof that Scott's perspective on the interplay between these different forms of knowledge was off the mark. He claimed with regard to the Bureau of Reclamation that he had found no evidence of the emergence of Scott's "modern authoritarian state" that was "obsessed with land and water planning, not just in the name of efficiency, but to justify (its) existence and maintain control over its subjects."³⁵ But Scott never claimed that expert knowledge *had to* result in high-modernist disaster. Indeed, he explained at some length that this was likely to occur only under certain specific circumstances, and he never precluded the coexistence of expert and more local forms of understanding, particularly if the former were prepared to adjust its practices to the specific situation at hand.

This melding of different forms of knowledge has been addressed by Tina Loo in the context of her own critique of Scott's work that builds on research into the construction of dams in British Columbia. In particular, Loo questions whether the engineers involved with such projects were in fact operating outside the "local." As she puts it, engineers, whom "Scott considers to be the exemplars of high modernism," were themselves involved with "knowledge [that] was in many ways local. It was characterized by an intense engagement with the biophysical world, which resulted in an understanding of it that was in part embodied, embedded in the particular, and characterized by an acceptance of limits." Loo refers to such knowledge as "high modernist local knowledge," expert knowledge that is grounded in the local, much as the MMRA engineers depicted in *Les aboiteaux* adjusted their procedures in line with local conditions. Engineers, from this perspective, were driven by "the impulse to control and dominate nature, on the one hand, and to live humbly and harmoniously with it on the other hand."[36]

A pertinent example of experts working sensitively within a particular local context is provided by yet another water-related project. The Prairie Farm Rehabilitation Administration (PFRA) was created by the Canadian government in 1935 to deal with the crisis faced by farmers dealing with drought conditions. While the MMRA dealt with too much water, the PFRA had to manage too little, and given this connection, it played a key role in setting up the agency in Atlantic Canada and subsequently in its initial operations. Indeed, the PFRA surfaces from time to time in the pages that follow as its engineers authored a number of the studies that set the stage for creating the MMRA.

In its first years, the PFRA focused on a range of irrigation problems. As Christopher Armstrong, Matthew Evenden, and H.V. Nelles have explained: "The agency's name emphasized rehabilitation: the sick land would be healed, partly with water." But the PFRA's actions did not extend to "fixing" the farmers, as was so common in other such rehabilitation projects, the agency choosing instead to work with the possessors of local knowledge, instructing them in new methods of land husbandry. Subsequently, the PFRA's mandate was extended in 1937 to include "community rehabilitation ... establish[ing] community pastures on the re-grassed lands of drought-stricken regions and transfer[ing] the displaced farming population to irrigated lands that could sustain them." There was no evidence of any coercion to attract farmers to participate in one of the first initiatives, the Rolling Hills resettlement project in Alberta, as there were more than 600 applications for fewer than 200 parcels of land. In order to ensure that it selected the "right" settlers, officials chose farmers who would bring appropriate knowledge and who "owned 'sufficient livestock and equipment in their own names ... free of encumbrances.'"[37] In this regard, the farmers' knowledge was seen as an asset, not a liability.

The PFRA's merging of local and expert knowledge was similarly evident with regard to the crisis faced by American farmers during the Depression. Jess Gilbert has written about a process that he called "low modernism," in which agricultural assistance was carried out by a central agency led by experts who refused to accept the essence of high modernism, namely, "the dismissal of local knowledge, history, tradition, and other 'illegible' activities like family farming." Instead, Gilbert found that there was a central role for "citizen participation" through a variety of shared activities, including "action research by technicians in partnership with rural communities."[38]

In the end, as Loo points out, historians have been too eager to "take high modernism as a given. Rather than being analyzed, it is applied as a label to connect particular events to a larger transnational historical moment."[39] She has amplified this point in her study of forced relocation in Canada, refusing to assume that individuals who were the object of government efforts to transform them were simply pawns of high-modernist schemes. Rather, she shows how such efforts, while "running roughshod over (certain) notions of freedom," also frequently resulted in "providing the poor with the means of empowerment, things like education and access to credit." In developing such policies, state bureaucrats employed expert knowledge, but when that didn't suffice, "the state also relied on expert knowledge of a different kind – that possessed by local people."[40] As both Loo and Gilbert show, the relationship between expert and local knowledge that could end up with the destructive imposition of high modernism, could also develop in other ways, displaying something close to Scott's description of mētis, locally based knowledge that was "open to innovation from below and outside."[41]

These studies looking at the interplay between local and expert knowledge, mostly in the context of projects connected to the movement of water, point to a variety of outcomes, largely dictated by the circumstances at hand. At one extreme, we see the transformation of the landscape and the people who inhabited it, classic examples of high modernism; and at the other end of the spectrum, cases where local and expert knowledge were successfully brought together. As we will see, the history of the MMRA offers the opportunity to explore various aspects of that interplay, bringing together marsh owners who had managed their environment without the intrusion of the state for roughly three centuries, and engineers who often spent their entire working careers in the region, developing internationally recognized expertise embedded in this particular tidal environment. In the process, these engineers became "locals" to some degree, adapting to local circumstances, as we see in *Les aboiteaux*. On other occasions, however, the MMRA's professionals ignored or demeaned the advice of those who interacted with the local environment without the pedigree of an engineering degree or participation in that professional community, resulting – with regard to the construction of tidal dams – in the sort of high-modernist disasters described by Scott.

Which Nature?

MMRA officials and marsh owners encountered one another in a variety of ways, depending on circumstances that changed significantly between the creation of the agency in 1948 and its transfer of control over the protective structures to the provinces in 1970. In order to make sense of these circumstances, the chapters that follow are built around the idea that a given landscape can be home to various forms of nature. In this particular case, there was first the marshland ecosystem that had existed for millennia, playing a central role in the lives of the Mi'kmaq First Nation and then very briefly for the Acadian settlers. In this first form of nature, the tides of the Bay of Fundy washed over and sustained the salt marshes twice daily, creating an environment that, as Matthew Hatvany has explained, "acre for acre, pound for pound ... produced more food in the form of fish and shellfish than fields of corn, wheat or other crops. On average, the salt marsh produces nearly ten tons of organic matter annually, in contrast to the best hay lands in North America which produce about four tons per year."[42]

The Acadians and their English-speaking successors transformed this environment, draining the marshes through the use of dykes and aboiteaux, in the process creating tens of thousands of acres of arable land that can best be understood by the term *dykeland*. In this context, there was some considerable irony in the fact that the MMRA's name suggested that it was in the business of rehabilitating *marshland,* as if the marshes that preceded the Europeans were going to return. After three centuries, the dyked landscape may have appeared "natural," but more accurately the MMRA's initial mission was to restore what William Cronon called "second nature ... the artificial nature that people erect atop the first nature."[43]

The first part of this book focuses on this stage in the agency's history, in the process building on the studies of authors such as Hatvany, who has explored the protection and expansion of second nature in the context of marshlands in the lower St. Lawrence region of Quebec. As Hatvany observed, the marsh owners of the Kamouraska area, much like those in the Bay of Fundy region, had long worked together to construct aboiteaux that drained their fields and provided additional arable land; in addition, as in New Brunswick and Nova Scotia, Hatvany found farmers who by the 1930s could no longer keep up with the repair of their aboiteaux using manual labour. They wanted to introduce heavy machinery but lacked the needed capital; and when they sought support from the Quebec government, it was limited compared with the massive investment by Ottawa in the MMRA, which was viewed as the gold standard by Aubert Hamel, a federal agronomist in the Kamouraska region. He observed in 1963 that "the example of the Bay of Fundy must stimulate us."[44]

Hamel wanted further government support for the protection of second nature, but by the time he made his appeal, the work of the MMRA in terms of repairing the dykes and aboiteaux had effectively been completed. The end of this phase in the MMRA's history was marked by the letting go of Ernie Partridge, one of its first employees, when he and his crew were no longer needed to make repairs. Over fifty years later, Ernie told me how crestfallen he had been with this part of his life coming to an end: "I was enthralled with my work ... I couldn't have had a better job than working with the tides."[45]

The agency carried out this stage of its mission with the encouragement, sometimes explicit and other times tacit, of the marsh owners, who needed to save their land from the tides in the face of crumbling protective structures. While the MMRA worked closely with some farmers who enthusiastically embraced the introduction of heavy machinery, its officials generally treated the proprietors as a group with a certain contempt, convinced that their practices had created the mess in the first place. Even though there were individuals, represented by Placide Landry in *Les aboiteaux*, who were drawn into the MMRA's operations, the marsh bodies as institutions ceased to play a significant role any longer. The MMRA engineers worked within the limits of the local environment, exhibiting "high-modernist local knowledge," at the same time showing a high-modernist disdain for the possessors of that knowledge.

Although the MMRA was largely done with the dykes and aboiteaux by the early 1960s, it continued to operate till the end of the decade, largely by busying itself with the construction of a series of tidal dams across five of the major rivers that flow into the Bay of Fundy. The agency believed that this practice, discussed at the very first meeting of the MMRA's Advisory Committee in 1949, would result in significant savings by blocking the movement of water upstream, in the process eliminating the need for the endless upkeep of the older protective structures, which could now simply be left to deteriorate. The MMRA engineers would be freed from having to deal any longer with the marsh owners, who were always viewed as a necessary evil, at best. Their local knowledge would now be irrelevant.

While the MMRA's work on the dykes and aboiteaux was restorative, the construction of the tidal dams was transformative, creating a "third nature," which provides the focus for the second part of this book. This transformation of the pertinent rivers was easy to see, marked by the creation of freshwater headponds upstream from the dams, and the buildup of silt downstream from the structures as sediment stirred up by the tides and previously distributed across the length of the rivers was now deposited on the seaward side of the dams. In an ironic twist, this sediment frequently provided an environment that permitted the emergence of new salt marshes, in a sense the return of "first nature."

Marsh owners had never expected the construction of such dams when they lobbied for protection of their lands in the 1940s, and in the first years of these projects, protests occurred as farming practices that involved access to the tides became impossible, making some local knowledge irrelevant. By and large, however, the marsh owners were content that the dams provided the protection they desired. But these dam projects created an entirely new context, not only for the marsh owners but also for the larger population in the region, which would have been relatively unconcerned by the rebuilding of the dykes and aboiteaux, which generally had little impact on those who weren't farmers.

Before the MMRA wound up its operations in 1970, its tidal dams were protecting over 32,000 acres of dykeland, roughly 40 percent of all the drained marsh that it was shielding from the tides. In addition, these dams made it possible to protect vast expanses of land that had nothing to do with agriculture, so that roads, railway lines, and buildings – both public and private – were now spared potential destruction from flooding; the dams were also instrumental in permitting the emergence of communities along the newly created freshwater headponds. On the other side of the ledger, however, the dams generated significant unease, particularly with regard to the marked decline in fish stocks in a number of the dammed rivers, as anadromous species such as salmon, which are born and which spawn in fresh water but spend most of their lives in salt water, found it difficult to negotiate the obstructions.

The costs and benefits of the MMRA's dam construction program were most clearly on display with regard to the one built across New Brunswick's Petitcodiac River. This project, completed in 1968, was never justified on the basis of protecting drained marshland, but rather was conceived as a solution to a transportation problem in Moncton, which required a new link to suburbs on the other side of the river. The MMRA might have constructed a bridge, but instead did what it was experienced in doing, and chose to block the river. In the process, fish stocks were destroyed, and the location of the dam far from the mouth of the river resulted in a spectacular buildup of silt that made the river downstream from the structure at Moncton significantly narrower and shallower. But none of this mattered as the MMRA and its partners were focused on carrying out this high-modernist project, reflecting what William Cronon described as the price "when human single-mindedness tempt[s] even the most enlightened managers into focusing their attention too narrowly on what most interests them."[46]

In this regard, the Petitcodiac Causeway can be viewed as part of a larger story of environmental destruction caused by the myopia of experts when it came to the control of water. Writing about the construction of a massive system of dams to divert rivers across the American West, Donald Worster has described the creation of a "hydraulic society" that was part of an "increasingly coercive, monolithic, and hierarchical system, ruled by a power elite based on the ownership of capital and expertise."[47] Focusing on the work of the US Bureau of Reclamation in the same region, Marc Reisner has written in a similar manner, noting how "we didn't have to build main-stem dams on rivers carrying vast loads of silt; we could have built more primitive offstream reservoirs ... but the federal engineers were enthralled by dams." Reisner concluded that "federal water development" was "a uniquely productive, creative vandalism ... The cost of this was a vandalization of both our natural heritage and our economic future, and the reckoning has not even begun."[48]

It is not necessary, however, to view the MMRA's dam projects, particularly the one across the Petitcodiac, in precisely such high-modernist terms. As we will see, the proposal to build the structure across the Petitcodiac was opposed by federal fisheries officers, who leaned on their own "high-modernist local knowledge" to anticipate the destruction of fish stocks. In accepting the advice of the MMRA's engineers and rejecting that of the fisheries officers, the Canadian government opted for a solution that favoured human convenience at the expense of the survival of another species. In the process, government officials were reflecting the anthropocentric thinking that had shaped the marsh environment since the time of the Acadians, a perspective facilitated by the conviction that marshlands were really wastelands. Nevertheless, the competition between highly trained professionals in the employ of the federal government complicates Scott's conception of the state, which was not a monolithic, muscle-bound entity with a single conception of what constituted expert knowledge.

If the views of fisheries officers did not carry the day in the 1960s, they did in the decades that followed as evidence of the damage done by the Petitcodiac Causeway accumulated, resulting in the river's being declared in the early 2000s as "the most endangered waterway in Canada."[49] This evidence formed part of a body of information stretching back to the 1950s that showed both the biodiversity of salt marshes and the challenges they were facing. In the process, a more ecocentric perspective on this environment emerged, brought to a non-scientific audience in 1969 by John and Mildred Teal's *Life and Death of the Salt Marsh*. The Teals explained how "salt marshes are valuable to us all. They are economically valuable in providing fish nursery grounds, protection from flood and storm damage, feeding grounds for birds, and sources of shellfish ... They are ... some of the most productive natural areas known."[50] Such evidence ultimately carried the day with regard to the Petitcodiac River, but only after a four-decade-long campaign that resulted in the opening of the structure's gates in 2010 and the decision in 2016 to remove much of the structure and replace it with a bridge.

In part, the length of the battle to restore the Petitcodiac was the result of opposition from the community that emerged along the headpond that was created, what residents who constructed lives there referred to as Lake Petitcodiac, so as to naturalize it. Alongside the destruction for which it was responsible, the Petitcodiac Causeway also created something new, reminding us, as Tina Loo has explained, that "we should pay attention to what is created as well as what is degraded or destroyed. In the case of high modernist projects such as dams, whole new communities came into being as a result of the 'disturbance' that came along with their construction."[51] Writing in this manner, Loo was rejecting the declensionist narrative that has had such a strong impact on the writing of environmental history. Responding to that narrative in slightly different terms, Mark Fiege has rejected the idea that "any human activity becomes just another story of ecological degradation." Instead, he has encouraged us to see "wildness" in all landscapes, seeing them as "hybrid landscapes, fusions of artifice and nature."[52]

In the end, the residents along the "lake" lost their fight as third nature was destroyed when the gates were opened and the headpond drained. But even with the opening of the gates and the restoration of fish stocks, the landscape that returned was not the one that predated human occupation, but rather a version that was every bit as manufactured as that marked by the headpond. The restoration of the Petitcodiac River effectively allowed the landscape to return to second nature, just as the MMRA's earlier efforts to restore the dykes and aboiteaux prevented first nature from returning. In the pages that follow, we will see how a given landscape can take on various forms of nature as a result of human interventions that were themselves the product of various forms of knowledge.

PART I
Second Nature

I
Out to Sea

DURING LATE SUMMER and early fall 1943, storms battered a large swath of New Brunswick and Nova Scotia, where dykes were holding back the tides of the Bay of Fundy, as they had for nearly three centuries. These structures had been in a weakened state for decades, prompting a high-level meeting of interested parties in early September in Amherst, Nova Scotia, a town in the heart of the marshland region that would soon become the site of the headquarters for the Maritime Marshland Rehabilitation Administration (MMRA). The meeting was chaired by E.S. Archibald, the director of the Dominion Experimental Farms, who plays a key role in the pages that follow.

After the committee had completed its work, Archibald took advantage of his presence in the region to provide a first-hand report on the damage done by the most recent storms, one of which saw "thirty-five hours of heavy, continuous rain."[1] He noted that

> a very large number of dykes had gone out in the last month with the excessively high tides and others may go out during this week especially if storms accompany the extraordinarily high tides which are prevailing. Already thousands of acres of good hay land have been flooded and the hay crop largely ruined for the next year or two.[2]

Archibald and some of his colleagues from the meeting "drove for the succeeding three days over large tracts of marshlands," concluding that while the current situation was bad, "it could be much worse" unless something was done: either repairing the dykes or eliminating them altogether "by blocking off tidal water in these areas, and putting a large aboiteau at the mouth of the stream."[3]

These were the two options that the MMRA would eventually adopt with vastly differing consequences, one restoring second nature, the other creating a new, third form through structures such as tidal dams. But in 1943, the ideas about how to deal with the problem had yet to be articulated very clearly. Instead, there was a certain amount of incredulity about how dykes and aboiteaux that seemed as if they had been part of the landscape forever were literally washing away, and with them drained marshland that had gone "out to sea," the expression that recurs in the correspondence from that era.

Archibald was a visitor to the region, but we also have testimony from people who had to live with the diminished dykes on a daily basis. One such individual was H.A. Francis, who in July 1943 (a few months before Archibald's visit) represented the views of the Marsh Owners' Association of Annapolis County before the Nova Scotia Royal Commission on Provincial Development and Rehabilitation, one of various inquiries set up at both the federal and provincial levels to explore how government could shape the postwar world. During its tour of the province, the Royal Commission made a number of stops where it looked into the vulnerable state of dykeland. In the case of a thirty-kilometre stretch on both sides of the Annapolis River, Francis said that there had once been 4,800 acres of dyked marsh. When asked how much of this dykeland had been lost, he thought hard and then said that "3000 acres, that is two-thirds has gone out to the tide and is absolutely useless. I could show you stretch after stretch ... that has gone out, and you will ask me why. It is owing to the fact of not keeping up with the dykes and aboiteaux."[4]

To provide the commissioners with a specific example, Francis pointed to Bloody Creek marsh: "It was a big marsh, but it is now nothing but a mud hole," because the aboiteau designed to stop water from entering and to allow standing water to flow out had washed away. "I talked to one of the dyke owners and I said, 'Is your aboiteau gone out again,' and he said, 'Yes, and we will let it stay out.' He said: 'Three years ago I spent $600 on the (aboiteau), and last year $400, and we have no aboiteau today.' So the water is coming in and it goes around behind the dyke and helps wash the dyke out."[5]

Nineteen forty-three was difficult for marsh owners, as the situation depicted in Figure 1.1 was repeated across the region. The Dykeland Rehabilitation Committee that Archibald helped set up reported in November

FIGURE 1.1 Middle Dyke Road flood, Grand-Pré, Nova Scotia, mid-1940s. *Acadia University Archives, Grand Pre Marsh Body Fonds, 2010.047-GMB/11b. Courtesy Ruth Conrad.*

that "large breaks in the dykes (have) occurred and in many cases, aboiteaux swept out to sea, resulting in the exceedingly heavy loss of hay and grain over very large tracts of dykeland." But the specific weather conditions, what the committee referred to as "exceptionally high tides and heavy rain and wind storms," would not have mattered if the structures had been in good repair.[6] After all, stormy weather was not unusual in this region, and yet the minute books of the marsh bodies, the local organizations that managed the protective structures, contain few reports of massive failures prior to the 1920s.[7]

No one really doubted that the weakened state of the dykes and aboiteaux was at the heart of the problems faced by farmers, but there was a significant difference of opinion over the role those same farmers played in creating the situation. On the one hand, the Dykeland Rehabilitation Committee recognized that the marsh owners were hard-pressed to make needed repairs "due to the shortage of labour and sufficient money."[8] On the other hand, various witnesses before the Nova Scotia Royal Commission referred to evidence of "neglect" on the part of the farmers, leading the commissioners to ask: "Even if the government should come to the help of the marsh owners and rehabilitate the dykes and aboiteaux and so on, what assurance is there that these will be maintained?"[9] As this question suggests, there was some concern that the local knowledge of the farmers was not up to the task at hand. In that context, this chapter explains how expert knowledge confronted local knowledge and how a system that largely functioned without direct government involvement for nearly three centuries became dependent on state intervention.

The End of First Nature

On their arrival in the seventeenth century, Acadian settlers entered a region with roughly 80,000 acres of salt marsh that had been formed over the course of millennia due to the tides of the Bay of Fundy, which – as much as the sun – governed the newcomers' lives. Twice daily, the tides make their way into the bay, moving up the various streams that flow into it, confounding the notion of what is upstream or downstream, all the while becoming larger and larger. By the time the tides reach the Minas Basin, near Grand-Pré, they can be up to fifteen metres in height. Sherman Bleakney, who has written extensively about this landscape, has observed that at mid-tide at this location, "the flow is equivalent to the combined flow of all rivers and streams on earth. From the sheer weight of that water, the land beneath the Minas Basin rhythmically sinks and rises with each tide. No other region on earth is quite like this one."[10]

Samuel de Champlain was typical of early European observers of Acadie, when he commented in 1604, on entering the Annapolis River: "From the mouth of the river to the point we reached are many meadows, but these are flooded at high tide."[11] The French newcomers were very conscious of the tides, but had little to say about the fauna and flora that thrived in this ecosystem thanks to the nutrients that washed up on the marshes and were trapped by its grasses. As George Matthiessen has explained, a salt marsh can produce ten tons of organic matter per acre per year, roughly what is produced via "terrestrial wheat production under modern and efficient methods of cultivation and is roughly equivalent to intensive sugar cane and rice production."[12] The Indigenous people of the region, the Mi'kmaq First Nation, understood this well, building their lives around the marshland "habitat for waterfowl and other animal life," and Acadian settlers followed their lead when they first arrived, taking advantage of the wild food resources available in the marshes. Ultimately, however, the Acadians worked at bringing first nature to an end, in the process aggravating relations between the Mi'kmaq and the newcomers.[13]

Within a year of commenting on the tides, Champlain was at work trying to remove unwanted water, in his words, "constructing a sluice against the seashore [at Port Royal], to drain off the water whenever I wished."[14] Decades after his brief stay in the region, there were efforts to draw profit from the marshes, this time by means of harvesting salt, but when this proved impractical, the French, leaning on their experience in Europe, turned to draining the marshes altogether to create fertile, arable land without the burden of clearing forests and removing rocks. In the process, they reflected a widespread understanding of the marsh environment that linked the natural buildup of sediment with the dyking of such soil. In this conception, the construction of dykes and aboiteaux became enmeshed in a natural process, leading to what Matthew Hatvany has called the "terrestrialization of wetlands," as "continual sedimentation and biological succession led to the steady advance of marshes, permitting successive reclamation."[15] Draining marshes simply sped up and improved on a natural process, at the same time responding to the sense that wetlands were chaotic places, neither land nor water, that needed to be tamed. Writing with regard to colonial America, Ann Vileisis has shown how turning this environment "into a physically ordered landscape was not only a religious obligation but a fundamental part of the colonists' worldview."[16]

There were other North American contexts in which European settlers drained salt marshes, but as Gregory Kennedy has explained: "The [Acadian] colonists' nearly exclusive reliance on marshlands for both immediate and long-term needs was distinctive."[17] As one French colonial governor observed: "They raise with so little labour large crops of hay, grain and flax, and feed such large herds of fine cattle that an easy means of subsistence is afforded, causing them altogether to neglect the rich upland."[18]

Although farming was largely restricted to the drained marshland, Acadians used their uplands for the construction of their farm buildings, a practice that created its own challenges for future occupants. As Yves Cormier has noted, in the nineteenth and early twentieth centuries, "farm owners would sometimes have to walk several kilometres in order to reach their drained land."[19] The division between marshland and upland also led to the construction of hay barns on the drained marshland, particularly in the vicinity of the Tantramar River in New Brunswick. Travelling through the region in 1885, Samuel Boardman observed "numberless barns to store the hay from harvest time until it is hauled to the home farms on the approach of the first snows. As many of the dyke lots are owned by farmers several miles away, it would be simply impossible for them to haul the hay home during the haying season."[20] In spite of these inconveniences, farmers saw the value of owning both types of land, and government intervention after the Second World War was justified, in part, to save farms that integrated dykeland and upland, each serving a distinctive function (see Figure 1.2).

By the time of the mid-eighteenth-century deportation of the Acadians, over 13,000 acres of the region's salt marshes had been drained, a process that was largely undertaken without any state intervention, a pattern that would continue until the creation of the MMRA.[21] As Kennedy has observed, "colonists left to their own devices carried out the drainage of marshlands," working as clusters of family units because

> a single family would have had difficulty building dykes and digging canals at the same time as gathering food, building shelter, and care-giving. A group of families, however, could combine their labour, completing the initial constructions required for the land to drain and begin desalination while also helping each other hunt, fish and gather, plant gardens, look after livestock and build homes.[22]

These familial clusters provided the basis for Acadian settlement under both French and (after 1713) British rule, developing areas that lent themselves to farming on drained marshland, mostly along rivers and streams that carried the tides of the Bay of Fundy. At first, the settler population was concentrated in the Annapolis Valley, near Port-Royal, the initial Acadian settlement in what would become Nova Scotia. In the late seventeenth and early eighteenth centuries, Acadians drained marshes in the area of Grand-Pré along the Minas Basin, and from there they extended their reach to the area of the Avon River (near present-day Windsor) and to that of the Salmon River (near present-day Truro).[23] During the final decades before the deportation, settlement expanded into what would become southeastern New Brunswick as salt marshes were drained along the Tantramar, Shepody, Memramcook, and Petitcodiac Rivers.[24]

In the centuries that followed, as dykes were moved further toward the sea to allow more land to be drained, the structures built by the Acadians sometimes turned up in the middle of farmers' fields. Boardman, writing in the late nineteenth century, could make out old dykes at Grand-Pré

FIGURE 1.2 Uplands and marshlands, with dykes in the background. *Original drawing by Caroline Boileau, 2019.*

that had "been plowed down and leveled off in places, but it is not a difficult matter to trace them."[25] In the early twenty-first century, I was shown by Dick Haliburton, a farmer from Avonport, not far from Grand-Pré, how the structures left over from the Acadian dykes provided the basis for a swimming hole in the midst of his family's farm, which had expanded through the construction of a dyke further out. Similarly, George Trueman, a farmer from southeastern New Brunswick, showed me a mound in the middle of his field, where the original Acadian dyke had been plowed over.[26] However, even if the extent of drained marshland increased after the expulsion of the Acadians, the marshland geography remained largely the same, with the MMRA eventually working in the same areas as had the Acadian settlers.

Given that there were no more than 14,000 Acadians in 1755, settling largely in the distinctive marsh areas, it was inevitable that there would be considerable distances between pockets of settlement. As Kennedy observed: "The distribution of these marshlands influenced the Acadians to adopt a more dispersed settlement pattern which gave them greater autonomy but also greater isolation."[27] Living on their own, the Acadians developed an unwillingness to recognize the authority of any state, either French or British, a practice that ultimately led to their forced removal by the latter.

In this context, the Acadians drained the marshes by means of a system many of whose characteristics remain unchanged to this day. In the process, they developed a set of skills, a form of local knowledge, regarding how to hold back the tides and make their lands suitable for cultivation. This process was led by an individual, chosen by the settlers and not by the state, known as *le sourd de marais* (the lord of the marsh). In research prepared for the film *Les aboiteaux*, Roméo LeBlanc, the future federal cabinet minister and governor general, described this individual as

> an elder in the community who was skilled in aboiteau construction. It was not at all uncommon to see the title passed on from generation to generation within the same family ... The "lord" would supervise the work, would study the water currents and the approaches to the sluice, as well as the direction of the tides and changes to the riverbed. He would decide on where to carry out the construction and nobody would ever think of challenging his experience and his knowledge in this regard.[28]

With the *sourd* directing the operation, teams of Acadians created and maintained three different elements of a system that, as a whole, turned salt marsh into land suitable for cultivation: dykes blocked the tides from entering the marsh; sluices known as aboiteaux were built into the dykes and were capped on the seaward side by a hinged gate (known as *le clapet*) that only opened out so that fresh water drained from the marsh at low tide, without allowing salt water to enter at high tide; and finally there was a system of ditches on the inland side of the dykes that channelled water to the aboiteau, to be expelled. Much like blood making its way to the heart, smaller ditches fed into ever larger ones, ultimately reaching a main channel that led the water out of the marsh (see Figures 1.3 and 1.4).

These elements needed to function together as a single unit because if any one part was compromised, the entire system would fail. For instance, if the ditches were not kept clear of obstructions, water would not flow from the fields, making it impossible for crops to grow. Similarly, if the hinged gate on the aboiteau was stuck open, salt water would be able to enter at high tide, removing the drained marsh from cultivation for several seasons, until it was again desalinated. The doomsday scenario saw the collapse of both the dykes and the aboiteaux, effectively allowing tides unobstructed access to the drained marsh, returning it to first nature, roughly the situation that E.S. Archibald observed first-hand in the fall of 1943.

Until the creation of the MMRA, most of the work involved in draining a marsh, and keeping it drained, was done by hand, following practices that were imported from France and then refined in the Atlantic Canadian mud. The Acadians built running dykes that "meandered along the high ground to each side of the creeks," which flowed into larger rivers and eventually the Bay of Fundy. The Acadians might have constructed barriers across streams, precisely what was done by the English-speakers who succeeded them after the deportation, but the Acadians, according to Sherman

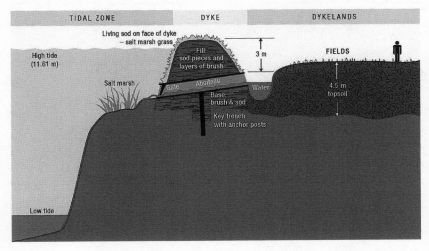

FIGURE 1.3 Cross-section of dykelands. While this graphic pertains to Grand-Pré, all dykelands were drained through this system. *Graphic by Steven Slipp. Image copyright Parks Canada. Reproduction based on an image produced by Parks Canada.*

Figure 1.4 Drainage ditches expel water from field. *Original drawing by Caroline Boileau, 2019.*

Bleakney, "had the foresight not to." Instead, they expended "a minimum of labour and materials ... The trade-off was that these miles of walls had to be patrolled and maintained, but this was easily accomplished."[29]

As for the construction of the dykes, Acadians assembled "teams of six men equipped with spades, pitch forks, and oxen. An assemblage of 120 men could construct more than a mile of dyke wall in just twenty days."[30] They began by installing anchor posts, around which they packed pieces of sod and marsh mud. This material anchored the dyke, around which walls were built out of bricks of sod that were carefully cut to a particular dimension so that they would fit together. The roots of the sod intermingled, creating walls that could stand up to the tides, the seaward side carefully sloped more than the inland to deflect the water crashing up against it.

The sod cutters were skilled craftspeople whose local knowledge changed little in the centuries between the arrival of the Acadians and the creation of the MMRA. In this regard, Ernie Partridge was a transitional figure who played a role in introducing machinery into the construction process but was also knowledgeable about how dykes had been built for centuries. Making a particular point of talking about the cutting and installation of the sod facing, Ernie explained how the starting point was to identify a sod pit relatively close to the work site, minimizing the time needed for transporting the cut sod by horse-drawn drags.[31] Time was also of the essence in cutting the sod, because workers further down the line would have nothing to do if the cutter did not produce, according to Sherman Bleakney, a 4-inch-by-8-inch brick every fifteen seconds. "As tide and time wait for no man, especially on marshlands, so it was that the production of the greatest number of uniform sod bricks in the shortest time was the ultimate key to quickly transforming saline tidal marsh into farmland."[32]

The crucial tool for carrying out the job was the dyking spade, a lightweight object with a sharp cutting edge that enabled it to slice through the roots. Sod cutters developed a close attachment to their spades, much as musicians to their instruments. When Bleakney interviewed Walter Kelly, a sod cutter then in his eighties, and handed him a spade, "his face lit up and his hands automatically grasped the shaft and twisted it as he showed the cutting motions." Bleakney also learned from his interviews that "the spade was never used for anything else for fear of dulling or nicking its edges. It always hung in a special place on the farm ... not to be touched by anyone other than the sod cutter."[33] Indeed, when I visited Ernie Partridge, a crucial part of his explanation of how dykes were built was a visit to his shed, where his spades were prominently displayed (see Figure 1.5).

Sod cutters wielded their spades as if they were extensions of their bodies. In this regard, there is the testimony of George Frail, whom we will meet again later as an engineer connected with several of the MMRA's dam projects. Frail was on the scene in the 1950s and so "was in the unique position of hiring the last of the professional dyke builders and of being able to stand aside and ... watch them perform." In an interview with Sherman Bleakney, he observed that the cutters "could produce sod while talking face to face with someone, not looking where they were cutting, somehow sensing what their spades were doing."[34] Their dexterity was similarly reflected in stories that Bleakney received from "the oldest dykers" he interviewed, who described lunch breaks that "involved contests of throwing sod the greatest distance, and wall builders who could catch a

FIGURE 1.5 Ernie Partridge and his dyking spades.

sod in mid-air on their fork, redirect it, and slam it into position in one continuous motion, with the precision of a bricklayer and the pride of a master craftsman."[35] And as was the case with the spade for sod cutting, the pitchforks were made particularly for the job, with "short, sturdy tines" so that the small holes produced would not "pry apart [the sod's] entanglement of roots."[36]

Like the dykes, the aboiteaux were the product of skilled craftspeople working in a particular environment, essentially what we see Placide doing in the film *Les aboiteaux*. We have the testimony of Jonathan Crane, who was "present [in 1764] when the first Aboiteau of any consequence was made here, by the English – which was superintended by two Frenchmen." Crane explained that it was important that "the bigness of the sluice ought to be in proportion to the fresh water that is to pass through it."[37]

If the sluice was too small, it would be impossible for water to drain properly, leaving some sitting in the fields; if it was built too large, limited resources would be wasted.

In some of the earliest Acadian dykes, sluices were constructed from hollowed-out tree trunks, but by the time Crane came on the scene, they were being built from "large hewn lumber joined well together, covered at the top and bottom with plank set into the timber and then sheathed with boards."[38] On the seaward side of the sluice, there was a clapper valve – a hinged gate made of wood in Acadian times – that allowed fresh water to leave the marsh at low tide but prevented new, salt water from entering. Over the course of two to three years, this process, aided by the careful construction of ditches on the inland side that channelled water toward the aboiteau, allowed the marsh to be drained and desalinated, permitting crops to be grown and cattle to graze.

With all the elements in place, first nature was replaced by second nature, marshes by arable land that was subsequently distributed among the local populace. While collective work allowed the drained marshland to exist, at least in the case of the dykeland at Grand-Pré, "collaboration did not mean collective ownership. Evidence suggests that once the collective transformation had been completed, the land was allotted through a lottery system. In order to consolidate fields or acquire better land, landowners would then trade or buy fields."[39]

Clearly, however, the process of consolidation was never entirely completed. A map of New Brunswick in the 1840s shows the random distribution of parcels of marsh among six proprietors, their properties having been distributed "by chance," so that there was "no logical order."[40] Even at the time the MMRA was created, small parcels of drained marsh still existed, distributed so that a farmer might own several that were not contiguous. For instance, a map that the agency created in 1951 for a marsh body along the Memramcook River, near the site of the filming of *Les aboiteaux*, indicated that one landholder, Elois Cormier, owned fourteen separate parcels, only four of which were contiguous with others that he owned; the rest were scattered across the dykeland.[41] The dispersed nature of land ownership underscored the point that individual farmers were inextricably linked with one another in the protection of their properties from the tides. At times, however, the communal management of the dykes and aboiteaux ran up against farmers' insistence that they had a right to manage their lands as they wished, a situation that we see in *Les aboiteaux* when the marsh owners at first refuse to assume their collective responsibilities.[42]

With drained marshland at their disposal, Acadians grew an array of grains, fruits, and vegetables, but the primary activity on the dykelands was the grazing of livestock. As Kennedy notes: "Where Acadian farmers surpassed most of their contemporaries was in the size of their herds. Already by 1671, there were more cattle than people in the colony, and almost as many sheep." This pattern continued into the eighteenth century, integrating the Acadians into export markets and making their new British masters nervous as "large numbers of cattle and sheep [were] being transported to French-held Louisbourg." More pragmatically, livestock appealed to marsh dwellers because "they were easier to transport than grain and could be moved if the dykes were breached by storms or raiders."[43] Centuries later the preoccupation with livestock – sometimes grazing on the dykeland, but more frequently housed on upland that was supplied by forage from the drained marsh – would lead the region's beef cattle interests to take the initiative in securing government support for dyke repairs, demands that ultimately resulted in creation of the MMRA.

WHILE THIS SECOND-NATURE landscape would continue intact well into the twentieth century, it would do so largely without the Acadians who had built it. Indeed, one of the reasons the British were so eager to gain control of this land was to make it available to American colonists, newcomers known as Planters, who took over the drained marshes on the removal of the Acadians. In early August 1755, only weeks after the deportation order had been issued, a letter was widely circulated in the colonial press that described the action as "a great and noble scheme ... If we effect their Expulsion, it will be one of the greatest Things that ever the English did in America; for by all accounts, that Part of the Country they possess, is as good Land as any in the World."[44] As Nova Scotia governor Charles Lawrence noted late in 1755, after the *grand dérangement* had begun: "As soon as the French are gone, I shall use my best endeavours to encourage People from the Continent to settle their lands ... and the additional circumstances of the Inhabitants evacuating the Country will, I flatter myself, greatly hasten this event, as it furnishes us with a large Quantity of good Land ready for immediate Cultivation."[45]

Among the Planters who came north to farm drained marshland was the Palmeter family, which arrived in Grand-Pré from Connecticut in the 1760s. Members of the family subsequently played a central role in the management of the marsh, and are still farming the same land over 250 years later.[46] But families such as the Palmeters did not exactly find the paradise they had been promised. For instance, at Grand-Pré they discovered in 1760 that much of the drained marshland was under water. With the Acadians no longer there to keep up with needed repairs, there was significant damage to the dykes in 1759 at the peak of the Saros cycle, when the relationship of the earth to both the sun and the moon results in exceptionally high tides every eighteen years.[47]

In spite of these problems, the newcomers appeared to have quickly acclimatized to marshland farming. New Englanders like the Palmeters may well have been familiar with what was involved in draining a marsh before moving north. As Kimberly Sebold has observed, "probate records and deeds suggest that reclamation occurred [in New England] during the colonial period."[48] And even if the Palmeters came without any previous experience, there were British settlers who could pass along their knowledge, having already developed expertise during the forty years between the conquest of Acadie and the deportation of the Acadians. Graeme Wynn has shown how "settlers used the marshlands in the vicinity of the British forts to supply the garrisons throughout the 1750s." Familiarity with marshland farming came from various directions, allowing Wynn to conclude that "the New Englanders soon put the marshland to use," with the result that "prices given for dyked marsh were consistently higher than those for undyked marsh and approximately the same as those for good upland in the late 1760s."[49]

In addition to knowledge secured within the British colonial world, the newcomers also benefited from the experience of the Acadians. Some of the former occupants of the land, now prisoners of war, were forced to "train" the new marsh owners. Jonathan Belcher, the chief justice of Nova Scotia, "asked military authorities to assist the Planters by distributing the Acadian prisoners [so that they] could be housed under guard and put to work maintaining the dykes." Belcher observed in 1761: "It appears extremely necessary that the [new] inhabitants should be assisted by the Acadians in repairing the dykes for the preservation and recovery of the marshlands."[50]

After the Acadians were allowed to return to Nova Scotia later in the 1760s, some were hired to provide paid labour to those who now farmed what had once been their land.[51] It was in this context that Jonathan Crane, whom we met earlier, was on hand as Acadians worked for some Planters in 1764. The direct Acadian contribution to the "new" local knowledge of the Planters can be heard even in the early twenty-first century by way of the terms that have migrated from one language to another. Aboiteaux are now widely referred to in English as "abideaux," and the sod blocks used to construct the dyke walls are called "parmang," from the Acadian *parment*.[52]

The only marshes available to Acadians as proprietors were at the northern reaches of pre-deportation drainage, in places such as Memramcook (in what would become New Brunswick in 1783), where the Planters had not taken over the land. As this remained a site where the Acadian connection with the drained landscape continued, it served as the backdrop for *Les aboiteaux*. Everywhere else, however, the marshes were now managed almost exclusively by English-speakers, their ranks bolstered in the 1770s by settlers from Yorkshire, "where fen-land was made productive by ditching, draining, and the construction of flood-gates." Much like the Planters, "it is quite possible that they applied their familiarity with wet-land reclamation in England to the problems of their new lands in Nova Scotia," and in the 1780s Loyalists became part of the mix.[53]

As during Acadian occupation of the drained marshland, the English-speaking proprietors looked after the dykes and aboiteaux with relatively little intervention by the state. While there were effectively no pertinent colonial statutes under earlier French and then British rule, this changed in 1760 when the Planters, faced with the inundation of their fields, received the support of the "provincial assembly [which] passed an act empowering proprietors to choose commissioners with the authority to repair and maintain the dykes."[54] This was essentially the system that prevailed until creation of the MMRA in the late 1940s.

The 1760 act allowed for the appointment of one or more "Commissioners of Sewers," adopting language from English statutes that referred to open drainage ditches, and not sewers in the more modern sense of the term. The preamble to the act noted that there were "great quantities of marsh, meadows, and low ground in this province, and particularly in the Bay of Fundy, and rivers, bays, and creeks, branching therefrom ... which by industry may be greatly improved."[55]

This improvement would come through the governor's appointment of commissioners "for the building and repairing such dykes ... as are necessary to prevent inundations." The commissioners would be named following "the request of the proprietors of such lands" in particular areas of the colony, in the process setting the stage for the division of the dykelands into marsh bodies, discrete groups of marsh owners, each of which was collectively working lands that required protection. The commissioners would be paid for their services and have the power to hire individuals to carry out needed work. The relevant costs would be covered by taxing marsh owners in relation "to each person's quantity of land and benefits

FIGURE 1.6 Grand Pre Marsh Body commemorative plaque.

to be received thereby." Proprietors could pay off their debt either through cash or by providing labour; if any failed to contribute, the shortfall would be covered by the other proprietors, who would take possession of the delinquent's lands pending reimbursement.[56]

In the nearly two centuries that followed this initial legislation, there were a variety of minor changes to the law that defined the operations of the dykelands, but nothing challenged the owners' autonomy.[57] This stability is reflected in the operations of several marsh bodies that can boast an uninterrupted corporate existence dating back to the immediate aftermath of the expulsion of the Acadians. Two such marsh bodies in the vicinity of Grand-Pré proclaimed their deep roots in the landscape in 1995 by erecting commemorative plaques on their 235th anniversary, with structures that were symbolically framed by timber from an old aboiteau and backed by the gate from the sluice (see Figure 1.6).[58]

Once settled in the region, the new proprietors set out to increase the extent of drained marshland. They did this in part by constructing much larger structures than the Acadians had ever built, an early example being their efforts in 1783 to install an "'aboideau' across the sixty-four-yard wide Missaguash River (which separates New Brunswick from Nova Scotia), an achievement which would have eclipsed, in magnitude, any Acadian dyke building." Such developments were closely connected to the growth of the local market, and so by the turn of the nineteenth century the drained marshland was largely "under grass and marked in the fall by 'vast stacks of hay made up in the true English manner.'"[59]

Hay remained the crop that fuelled the local economy until the early twentieth century, some sold on the open market and some used to feed cattle, both beef and dairy. While it is commonplace to speak of the difficulties of the Atlantic Canadian economy, this did not apply to the proprietors of drained marsh, who, as long as hay prices remained strong, were able to maintain their dykes, aboiteaux, and ditches, and to regularly expand their holdings by protecting ever more land.

The stability of the marsh owners' affairs was reflected, for instance, in the annual meetings of the marsh bodies. Typical of the minutes I was able to locate were those for the Dentiballis Marsh in the Annapolis Valley, which spoke to a routine that remained unchanged for decades. From the 1880s to the 1940s, annual meetings were held in early September so that dates could be set to allow cattle onto the marsh for grazing and for their removal, with someone appointed along the way to look after branding so that only the owners benefited. As George Warren, a visitor to the region, noted in 1911: "Hay was typically cut between the middle and the latter part of July, and about September 10, thousands of cattle were turned onto the marshes to feed and fatten until November."[60] Year after year, the Dentiballis minutes, and those of other marsh bodies, formulaically reported the setting of those dates, the approval of the marsh body's finances, and the provision of funds for minor repairs.

A sense of stability was also conveyed by several American visitors to the region, who reported on the drained marshes as part of studies designed to improve the management of salt marshes in the United States. For instance, Samuel Boardman, who journeyed there in 1885, wrote positively about the area near the Dentiballis Marsh:

> All along the valley of the Annapolis the dyked lands are among the most valuable of the improved lands, and are wholly devoted to grass and grain ... Dyked land of good quality is worth here $150 to $200 per acre, and in

some instances even $400. Such land will yield from 2½ to 4 tons of the best timothy [hay] and clover, per acre.[61]

Even more enthusiastic were the reports of George Warren, who visited the region twenty-five years after Boardman. Wherever he travelled, he wrote excitedly about how "many tracts [of drained marsh] have been cropped for generations without renovations of any kind, and without once failing to yield good crops of the best English grasses." The situation was so positive that "not only farmers, but merchants and professional men who are looking for good, sound 8 and 10 percent investments believe and invest in these reclaimed marshlands. The leading social club of the thriving city of Amherst is named the Marshlands Club." Writing for an American audience, Warren was so impressed with what he found that he was left to wonder how the Canadian provinces "have made such pronounced success in marsh-reclamation work, while the great bulk of our [American] marshes ... are still in their natural state, or where reclaimed the successes have been few or indifferent."[62] Warren could have hardly anticipated that by the 1940s some of this dykeland would be washing out to sea.

The Experts Arrive

During the early decades of the twentieth century, it appeared that first nature had been tamed, marshes having been transformed into drained, arable, and productive land. As in many other parts of the world, here was the seemingly "natural terrestrialization of wetlands."[63] To be sure, there were moments when even the best efforts of proprietors were no match for the elements, as during the 1869 Saxby Gale, which resulted in the highest tides ever recorded in the Bay of Fundy. George Warren observed how the tide "rose in different parts of the bay from four to eight feet above high-water springs ... submerging the marsh lands generally."[64]

By and large, however, whatever challenges arose appear to have been handled successfully through the mobilization of local resources. This positive depiction of local efforts to maintain second nature was reinforced by the photographic record, which frequently showed workers successfully fixing problems connected with their dykes through the use of manual labour, as had been the case for centuries (see, for example, Figure 1.7). This local control, however, depended on the marsh owners sustaining an economy largely based on raising hay on the drained marshland. Hay was

FIGURE 1.7 Work crew on the Wallace Bay aboiteau, ca. 1910. *Courtesy North Cumberland Historical Society.*

FIGURE 1.8 Hay barn on the Tantramar Marsh, New Brunswick. *Courtesy of Shaun Cunningham.*

so dominant in production on the drained marshes of the Tantramar region in southeastern New Brunswick that the region was commonly referred to as the "World's Largest Hayfield," its iconic hay barns defining the landscape (see Figure 1.8).

This dependence on hay was nothing new, but in the late nineteenth and early twentieth centuries, many farmers moved away from the sale of beef cattle, fattened on their hay. W.W. Baird, the superintendent of the experimental farm in Nappan, Nova Scotia, and an important figure in developing policy to assist farmers when their dykes began to deteriorate, observed that there had once been "annual shipments out of the Amherst and Sackville districts of 4,000 to 5,000 heads." However, these shipments began to decline at the turn of the century, and seemed unlikely to return because "the beef industry was being developed on the prairies and keen competition was soon felt from western beef."[65]

Farmers adjusted to the new circumstances, happily selling their hay on the market in the face of skyrocketing prices. Warren found hay prices exceeding $10 per ton in 1911, peaking at nearly $30 in 1918.[66] In this context, the Special Beef Cattle Committee, created to help farmers on drained marsh return to the sale of cattle, reported in the late 1930s that "these farmers are *hay-minded* and when hay sells for *seven dollars per ton or better*, the major portion of the hay is sold and cattle are curtailed in number."[67]

Heavily reliant on the hay market, farmers across the region were devastated when prices began to plummet, falling to $23 in 1923 and to $10 a decade later, before bottoming out at around $6 per ton by the late 1930s.[68] This situation was no doubt exacerbated by the Depression, but it began earlier and was directly related to the declining market for hay that was no longer in demand for feeding horses, which were being displaced by vehicles powered by the internal combustion engine. In Canada as a whole, the number of horses reached an all-time high of 3.5 million in 1921, falling by nearly 20 percent over the next two decades. This decline was evident in urban areas, where streetcars and automobiles became dominant, as well as in rural areas, where the tractor and the truck displaced the horse.[69] As the Special Beef Cattle Committee observed, "the replacement of the horse with other types of power had noticeably lessened the volume of market for good horse hay, which was a valuable market for many of these growers."[70]

By the late 1930s, farmers were struggling with much-reduced incomes that could barely cover their expenses. According to Baird, "this was the end for the hay farmer, for after deducting $2.50 for cutting, $1.50 for pressing and fifty cents per ton for cartage, it left nothing for interest or maintenance charges."[71] To make matters worse, at the same time that prices were collapsing, farmers' yields were also in decline, one report noting in 1940 that "there are many such areas where the crop is less than half its former yield." In light of all the problems in the region, "land values have likewise been affected. Land that formerly sold readily at $110 to $150 per acre is now a drag on the market at twenty dollars to thirty dollars per acre."[72]

The collapse of the hay economy led directly to the deterioration of drained marshland as owners, their incomes sapped, found it difficult to maintain either their protective structures or the drainage channels on their lands. As a result, by the late 1930s nearly 75,000 acres of drained marsh, out of a total of 80,000 across the two provinces, were in danger of reverting to first nature.[73] Faced with this unprecedented crisis, the dykelands were studied as never before. Long managed by proprietors through their marsh bodies, the area became a magnet in the late 1930s and early 1940s for organizations and individuals dedicated to applying professional expertise to fix the problems. By and large, those who came to study the region were connected with the federal, New Brunswick, or Nova Scotia governments; they came from as far away as Saskatchewan, and arrived with long years of experience, often grounded in engineering or through association with Canada's chain of experimental farms. They did not always agree with each other on how the dykelands should be fixed, but they were all confident that their expertise provided them with informed solutions to a difficult situation.

In this context, three separate inquiries in 1939 and 1940 alone focused on dykelands on either side of the Nova Scotia–New Brunswick border, frequently employing questionnaires distributed among the farmers to arrive at results through the scientific, methodical collection of data. The Special Beef Cattle Committee's 1939 study of the region, supplemented by a further inquiry the following year, was based on a survey of nearly 1,200 farms. To dig deeper into the situation, the committee carried out a "detailed study of 280 farms in the Sackville and Westmorland parishes" of New Brunswick, together with "a carefully prepared questionnaire [that] was sent out to some seventy odd farmers in the Cumberland district in Nova Scotia."[74] Cumberland County was also mined for data by J.E. Lattimer, an economist from McGill University's agricultural college, who spent the summer of 1940 collecting information from 128 farms.[75] The most elaborate project was the Nova Scotia Marsh Survey, which – in spite of a name suggesting a broader mandate – was also focused on Cumberland County.[76]

To carry out the Marsh Survey, J.W. Byers, from Nova Scotia's agricultural engineering office, was accompanied by five student assistants during the summer of 1939. Byers explained: "As the summer was unusually dry the greater part of time was spent in [carrying out] the field measurements; rainy days were spent in calculations." In addition, "questionnaires were sent out to land owners and to Commissioners of Sewers ... As the party was organized, all the members were needed to keep the work going ahead, and there was no opportunity for one man to be free." Byers seemed to be self-conscious about the lengths to which he and his team went to carry out their survey: "It might be said that there were more pains taken in making [the survey] than the project justified," but for "a project of the magnitude of the Marsh Survey, [it] is worth taking special precautions [so] that the greatest possible usefulness may be obtained from it."[77]

These studies collectively provide a vivid picture of the situation facing farmers at the time, paying considerable attention to the deterioration of the dykes. For instance, the Special Beef Cattle Committee described how "during recent years, dykes for miles have been undermined and washed in over the growing crop and some have slipped off into the sea, while others have been left an easy prey to high tides, when accompanied by strong winds."[78] For its part, the Nova Scotia Marsh Survey reported that aside from 15 to 20 percent of the region where the "dykes were in fairly good condition ... the remainder of the area is in very poor condition ... Dykes are battered and weak; aboiteaux are battered and leaking. There are all stages of deterioration, but this is a fair picture of the marshes."[79]

But while the washing away of the dykes might have been the most visible and dramatic part of the crisis, the challenge for marsh owners extended to their fields, where poor drainage often made it impossible for water to reach the aboiteaux so that it could be expelled at low tide. The Special Beef Cattle Committee report noted that with the deterioration of the dykes, there were "heavy losses to the growers, for not only do they lose their crop, but the drainage systems are filled with sediment each time the tide entered."[80] The Nova Scotia Marsh Survey pointed to one particularly disastrous case, describing drainage at the Amherst Marsh as "*Extremely unsatisfactory.*" The survey team had trouble getting out into the marsh "on account of the quantities of water lying [there] after at least a month of dry weather." They found "6 inches of water all over the marsh. One of the party ... waded in water up to the waist, and went through an old ruined barn which was practically floating off its foundations." In response to the survey's questionnaire, one marsh owner observed: "This marsh drains a large area of upland and after a heavy rain I have seen the marsh in parts flooded from one to two feet deep which stayed there until the hay and grass were completely ruined."[81]

In addition to describing the problems that farmers were facing, the studies also assessed the roots of the crisis. Dependent on manual labour to both keep up the dykes and aboiteaux and maintain the ditches in the fields, farmers were faced with ever-rising costs at the same time that their incomes were in decline. The Special Beef Cattle Committee focused on the "yearly decrease in available, experienced dykers and ditchers. Most of this work has been done, in the past, by spade work. This naturally brings up the question of experienced help or efficient machinery for this kind of work."[82]

In terms of finding help, farmers were hamstrung by the exodus of young people who were leaving to find better prospects elsewhere, a common situation across rural Atlantic Canada at the time. In his 1939 survey of the Tantramar Marsh, H.R. Hare told the story of one farmer who was "greatly concerned over one of his four sons who was associated with him on the farm. The three other sons had gone out into the world and were doing well, coming back weekends with a nice car, while his boy at home found it necessary to work Sundays and all, and could not afford to own a car or get married."[83] Along the same lines, there is a poignant scene in *Les aboiteaux* in which Placide, the *sourd de marais,* is boarding a bus to check out the MMRA facilities in Amherst. At the bus stop, he comes across a group of young men waiting for transport to take them to work in the city. In the absence of such labour, farmers were faced with inflated costs for those who remained, leading one farmer in the film to observe: "You've got to understand, Placide, a day's wage is too much money. These days, we can't get ahead by paying men to work on the dykes."[84]

As for the use of machinery, the time had not yet come when this was a real option for most farmers. To be sure, by the 1930s there were already exceptional occasions on which equipment such as draglines along with bulldozers were being used, both in the Bay of Fundy region and in other dykeland contexts such as the Kamouraska region of Quebec. A dragline, like the one from the early 1950s shown in Figure 1.9, was essentially a crane with a large bucket on the end that, through a system of cables, was able to excavate in minutes what would take men days to accomplish. Matthew Hatvany has described the introduction of such machinery in Kamouraska in 1938, enabling the protection of over 800 acres.[85] A year later, a dragline was used for the first time in Atlantic Canada as the Grand Pre Marsh Body found itself in a bind when a section of dyke on the easterly side of the marsh was gradually losing its protective foreshore. The danger of a break in the dyke became serious. At that time, it was impossible to get men to rebuild dyke according to the traditional method. R.H. Palmeter, the commissioner, arranged for the first trial of a large dragline-built dyke on the Maritime Marshlands.[86]

In both of these situations, the farmers financed the use of machinery on their own, without government support. As a result, the Kamouraska experiment appears to have been a one-off, while the Grand-Pré initiative was the exception that proved the rule, as its farmers were relatively well-off, enabling them to introduce an innovation beyond the means of most marsh bodies. As part of their 1950 study of about 700 farms dependent on drained marsh, Gordon Haase and D.J. Packman pointed to the distinctive character of agriculture at Grand-Pré, where dairy farming shielded marsh owners from the decline in the demand for hay. As Haase and Packman put it, "the greatest concentration of fluid milk producers was in this area," leading to an "intensity of farming operations that [was] reflected in the very high investment on the farms studied."[87] By contrast, they found in other areas such as the Tantramar region of southeastern New Brunswick that a lack of resources resulted in "cultivation [that] was less intense."[88]

FIGURE 1.9 Dragline at John Lusby Marsh, early 1950s. *NSDA, Nappan, NS. Reproduced with permission of Nova Scotia Department of Agriculture.*

With the decline of the hay economy, farmers had little incentive to improve their fields, and marsh bodies were hard-pressed to fund improvements as urban residents and absentee farmers, who had acquired land in the times of strong demand for hay, now abandoned their properties, so that their "failure to pay the assessment is fairly general."[89] The widespread shortage of funds led the Special Beef Cattle Committee to recommend government assistance to give farmers access to "the best and most suitable type or types of machinery to be used for dyking and ditching."[90]

Beyond the question of funds, farmers might also have thought twice about investing in machinery, since the results at Grand-Pré were mixed at best. An MMRA report from 1949 indicated that the 1939 dragline experiment had

> appeared very satisfactory, when the job was done. It was hoped that it would soon sod over and withstand erosion as well as the traditional sod-faced dykes. A few years proved otherwise. It was found necessary to erect plank facing along the most exposed end. The remainder was later faced with brush and stakes. On that exposed shore, it was questionable if the brush and stake work was really economical.[91]

Grand-Pré had a similar experience in 1944 with a dyke that was also built with a dragline, following "the same general procedure as in 1939 ... On the first impression, one might easily have said, 'It's a good job, very uniform and neat.'" Within a few years, however, it became clear that "in an exposed location, the dragline dyke could not stand up."[92]

In the late 1930s and early 1940s, as the experts made their way through the dykelands, marsh owners were understandably reluctant to embrace new dykeland management practices, which were both beyond their means and still unproven. Nevertheless, those same experts often took the view that it was the farmers' unwillingness to abandon the traditional way of doing things, their local knowledge, that formed a significant part of the problem. For instance, on the issue of dyke building, J.W. Byers of the Nova Scotia Marsh Survey collected evidence from farmers who were convinced that dykes had to be constructed using centuries-old practices. They explained that it was necessary to "build a dyke in layers, each layer well tamped and packed before another layer is placed ... The whole process is one of hand labour and requires some genuinely skilled labour and good supervision." He listened patiently to the marsh owners' reservations about the use of dyking machines but dismissed them because "the only comments obtained were unfavourable." While the experience at Grand-Pré had shown that there were good reasons for such resistance to change, Byers was sure, although it was little more than an article of faith at the time, that there had to be "some method of building dyke by the use of a power shovel which would be cheaper and more satisfactory than the method of hand labour."[93]

More broadly, Byers was convinced that the farmers' stubborn adherence to outdated practices was an obstacle to the rehabilitation of the dykelands. To be sure, he recognized that the farmers had been stuck with low incomes and high maintenance costs, leading some to conclude that "the marsh was more of a white elephant than an asset." At the same time, however, he was convinced that blame also needed to be directed toward, what he called the "Attitude of the Marsh Land Owners," an expression that was given prominence as the title to a separate section of his report. While Byers described some farmers as having been open to collaboration with his team, he viewed many others as resistant to any outside intrusion, pointing to occasions when "members of the [Marsh Survey] party were regarded as trespassers on private property." Moreover, Byers observed that there were farmers who, when faced with deteriorating infrastructure, were ready to accept financial assistance as long as it came "without the advice of anybody." Putting it quite baldly, he characterized farmers as having shown "a lack of interest [with] plenty of idle time in which they could drain and take care of their marsh."[94]

However, the single issue that most clearly showed the reluctance of some experts to trust the farmer's local knowledge pertained to the centuries-old practice of tiding (sometimes called warping, flowing, or drowning), which saw marsh owners flooding their lands from time to time so that silt might be deposited to add to the soil's fertility. They did this either by raising the gates on the aboiteaux at high tide or by creating a breach in the dyke to allow the tides to rush in. Farmers recognized that the practice was not without its costs, as the fields that had been tided with salt water would be unavailable for cultivation while the salt leached out. The Canadian naturalist W.F. Ganong observed that some proprietors did not engage in tiding because they "cannot afford to lose all return from their land for several years." Nevertheless, he thought that this reluctance was short-sighted because of the fertility that the new mud brought to the land.[95]

The Acadians engaged in tiding, as did newcomers from Yorkshire who were already familiar with the practice when they arrived in the region after the deportation.[96] Tiding was still going strong in the late nineteenth century, when J.R. Sheldon observed how tides were allowed to flow across the fields to "deposit a coating of finely granulated mud, which serves as

FIGURE 1.10 Repairing aboiteau after tiding, Falmouth Village Great Dyke Marsh Body, 1906. *Courtesy Philip Davison, Falmouth, NS.*

a dressing of the best possible manure, and operates for many years in this capacity."[97] More specifically, the practice was followed at the experimental farm at Nappan, Nova Scotia, that had been designed to serve as a model for farmers across the region. For decades after the farm's creation in 1886, "the land continued to produce good hay crops while the practice of opening the flood gates periodically was followed."[98] Tiding was also employed in 1906 after the practice had been approved, at the first meeting of the Falmouth Village Great Dyke Marsh Body, by proprietors with land along Nova Scotia's Avon River who wanted to receive "the benefit from the cutting of the two Aboiteaux and flowing of the said Dyke [i.e., drained marsh]." Figure 1.10 shows the marsh owners reinstalling the aboiteau after the tiding operation. For his part, George Warren, the American whose report we have already seen, described in 1911 how the widespread use of tiding had "resulted in the creation of thousands of acres of fine land."[99]

More recently, Harry Thurston has explained how a neighbour on dykeland not far from Amherst "would [beginning in the 1950s] remove the clapper on the aboiteau to allow the tides to flood his hay field ... renewing the fertility of the soil."[100] This was a mainstream practice that figured in all provincial statutes connected with the management of marshes, starting with the introduction of the first such acts in 1760 and continuing until the establishment of the MMRA nearly two centuries later.[101]

Tiding was so prevalent that the Nova Scotia Marsh Survey made a point of collecting evidence on the question from farmers in the late 1930s; not surprisingly, it found that "the general opinion is that tiding is beneficial, either for raising the level of the marsh or for enriching the soil." Fairly typical was the McGowan Marsh Body (in the vicinity of Amherst), which reported: "After marsh has been overflowed the land is firmer to work on. The ditches and cross drains do not crumble and cave in so quickly. The hay is much better ... Tiding also kills weeds and helps to level the land, making it easier to drain." Farmers differed, however, as to whether opening the aboiteau (as opposed to creating a breach in the dyke) was the best way to achieve the desired end. The Embree Marsh Body (adjacent to the McGowan Marsh) argued that raising the gates on the aboiteaux did not allow "volume enough of water to come in."[102]

Very few of the reporting marsh bodies categorically responded that tiding as a procedure should be abandoned, but one that did received high marks from the Nova Scotia Marsh Survey for eschewing the practice. The Barronsfield Marsh Body, also in the vicinity of Amherst, was described positively in terms of the upkeep of its aboiteaux and the quality of its drainage. In his report, Byers tellingly concluded that "this marsh is among the best surveyed, which explains the owners' objection to tiding." More generally, Byers viewed tiding with skepticism. He recognized that "it is commonly held that there is great advantage to be had from allowing the tide to flow in over the marshes and so to enrich the soil by deposit of fresh mud."[103] From his professional perspective, however, there were problems: "Broadleaf marsh can be improved by mud deposit, but if the deposit is too heavy there will be a loss of crop for a year or two. On the English hay marsh, the salt water will kill the hay and make plowing and re-seeding necessary, so the owners of such marshes do not believe in it."[104]

By far, the strongest opposition to tiding came from E.S. Archibald, the director of the Dominion Experimental Farms, whom we met at the beginning of this chapter during his 1943 tour of the dykelands. Born in Nova Scotia in 1885, Archibald quickly rose through the ranks of the experimental farms, becoming its head in 1919, a position that he held until

1951. Interested in bringing science to Canadian agriculture, he "raised the educational standards of experimental farm staffs by hiring scientists with advanced degrees and by promoting educational leave for others."[105] During his leadership, he played a key role in establishing the Prairie Farm Rehabilitation Administration (PFRA) in 1935, an agency that he headed for its first years and that, as we will see, provided both a model and personnel for the MMRA.[106]

Archibald would be the single most important individual associated with the marshlands dossier in Ottawa during the 1940s. In this context, he minced no words in his correspondence about his low regard for the marsh owners' practices, writing at the start of the decade to G.S.H. Barton, his deputy minister, about how "the methods used in handling dykes in Nova Scotia and New Brunswick are still very primitive."[107] More specifically, he could not understand the persistent use of tiding, and so reacted with horror when it came to his attention in 1940 that W.D. Davies, a senior official with the federal Department of Agriculture, who was in the region along with many others to explore how beef cattle might be reintroduced, had reported positively about the practice. Davies observed: "Where the dykes have been broken and a new sediment has been deposited by the tide the quality of the hay is very good."[108] In response, Archibald wrote to Barton, wondering whether "Mr. Davies had been reading Longfellow's *Evangeline*, in which it is stated that at certain seasons the dykes were opened to allow the tides to wander at will over the meadows."[109] More precisely, Longfellow wrote:

> Dikes, that the hands of farmers had raised with labor incessant,
> Shut out the turbulent tides; but at stated seasons the flood-gates
> Opened, and welcomed the sea to wander at will o'er the meadows[110]

Longfellow's epic poem, which romanticized Acadian life before the deportation, was the antithesis of the new evidence-driven approach championed by the experts examining dykeland farming practices. In this context, Archibald's choice of this particular device to criticize Davies, and by extension the farmers, was a harsh rebuke. But Archibald went further, taking Davies to task for being so misinformed:

> In the early days of building these marsh lands, [tiding] might have been the practice, but those of us who have had to handle dyke lands which have been flooded by tides because of broken aboiteaux and dykes realize that the saline deposit of mud from the tides just makes it impossible to do much

in the way of bringing the land back to hay or crop production for a period of two years or more.[111]

As we have seen, W.F. Ganong recognized that the intrusion of salt into farmers' fields was a real problem. He suggested, however, that damage to the soil could be reduced "by admitting only a little tide at a time, or by admitting it only in late autumn after the ground is frozen, when the grasses are little injured by it."[112] Indeed, nearly all of the respondents to the Nova Scotia Marsh Survey reported that tiding should take place in the fall or in the spring, "preferably when lands are wet so it will not salt badly." These farmers indicated that tiding was not some folk practice that was employed arbitrarily, but this did not impress Archibald, who maintained that farmers could best "raise their soil fertility through the use of chemical fertilizers and legumes."[113]

Closely related to the question of tiding, Archibald also advocated for a radical rethinking of where aboiteaux should be constructed. In his report of the findings of the Marsh Survey, Byers estimated that each aboiteau cost at least $2,000, no small sum for marsh owners with limited means, and he went on to observe that "there is a question on whether it is advisable to run dykes along the sides of a creek and have a small aboiteau well up the stream, or to have a large aboiteau near the mouth of the creek and so gain more land along the side of the creek."[114] In the latter case, fewer aboiteaux would be required as entire creeks, or even rivers, could be cut off from the tides, making the dykes and aboiteaux upstream superfluous. In the process, tiding would become impossible as marsh owners would no longer be able to open specific aboiteaux or sections of dykes to allow the tides to enter. Archibald was the first of the experts to weigh in on this matter, in 1942 encouraging Barton to consider "the construction of dams across drainage channels [that] might save many miles of maintenance of dykes. These are engineering jobs which I think require pretty special study."[115] A few weeks later, writing once again to his deputy minister, Archibald reinforced his conviction that the tides could be mastered through application of the principles of "modern engineering."[116]

By consistently vaunting the application of scientific knowledge while denigrating the practices of farmers, Archibald represented what James Scott had in mind when he referred to experts who "regarded themselves as much smarter and farseeing than they really were and, at the same time, regarded their subjects as far more stupid and incompetent than *they* really were."[117] Of course, experts did not *have* to dismiss local knowledge. Indeed, Scott pointed to the ideal of integrating expert and more experiential knowledge, what he called "mētis." For her part, Tina Loo referred to the shaping of expert knowledge to accommodate local circumstances as "high modernist local knowledge."

In the case of the dykelands, we have an example of the sort of thinking that Loo had in mind in the assessment of the situation in 1943 by Ben Russell, at the time a senior consulting engineer for the PFRA, where he had developed expertise in "the construction of stock-watering dams and garden-watering dugouts on individual farms within the dust bowl area."[118] Archibald wanted to consult with someone who had "wide experience in drainage and irrigation," and so he brought Russell to the region, perhaps expecting confirmation for his view; but if that was his hope, he was sorely disappointed.[119]

Russell came away from his trip east with considerable respect for the marsh owners, particularly those whom he described as the "old time practical dyke men," who had distinguished themselves through their

> skill and hard work ... All of the works, such as dykes, aboideaux, and drainage ditches were designed and constructed without any assistance from engineers, and ... with little knowledge of the conditions to be met such as freshwater discharge, reservoir capacity ... The wonder is that the works have so long performed the service for which they were constructed. All respect is due to these old builders for the excellent service they performed.[120]

This respect did not prevent Russell from suggesting better ways of operating when they presented themselves. For instance, he was critical of the marsh owners for resisting the use of modern tools "such as topographical surveys, aerial photographs, and other necessary data." Russell recognized that "it is quite possible with the local knowledge of the individual parcels to put in drainage systems that will carry off surplus waters without the aid of any comprehensive surveys, [but] it is not possible to locate the most economical and best systems of drainage or to determine the best location of dykes, aboideaux, and other structures, without such maps."[121] In this case, he was not suggesting the abandonment of long-standing practices, but rather their improvement through the integration of new technology within existing local practices, which was at the heart of mētis.

In terms of tiding, Russell listened carefully to various perspectives. The marsh owners told him how "some of the marshes have deteriorated and require tiding to refertilize the soil." As for the experts, referred to as "agriculturalists," Russell observed that they "did not agree with [the farmers' views], and are inclined to think that good drainage is all that is necessary to bring [the fields] back into production." Russell recognized that "it will be difficult to convince the present owners that tiding is not the most economical and best means of restoring the fertility of these poorer lands." In discussions with dyke commissioners, he learned how the tide could be let onto the fields in the fall with little risk of damage from the salt, leading him to conclude that farmers should be encouraged to continue the practice, if they were so inclined. In fact, he went so far as to recommend that "in the design of aboideaux in future the gates should be large enough and so designed as to be readily removed or raised in a frame to let the tide go by regulated as required."[122]

Russell also parted ways with Archibald in terms of the construction of structures across the region's rivers, which would have ended tiding. Once again, he listened to people on the ground, in this case those involved with experiments concerning tiding who concluded that "it is a mistake to permanently shut off the sea and that every means should be provided so that the tides can flow back and forth through the main streams and drainage channels."[123]

In a sense, Russell's report provides a glimpse of what might have been if the experts in the region had more carefully listened to the marsh owners instead of dismissing them. Perhaps Russell was able to look at them differently because he was not from the area, and had not developed fixed views about the farmers. In any event, he returned to the Prairies, leaving the views of individuals such as Archibald to dominate. Convinced that they had nothing to learn from the marsh owners, these experts held sway as government intervention in the marshlands began to take shape.

Plugging the Holes in the Dykes

The various expert studies were preliminary to governments (both federal and provincial) taking action to *rehabilitate* the marshlands. This expression was in widespread use in numerous contexts in various national settings at the time, often applied in reference to the transformation of both specific environments and the people who inhabited them. In this particular case, what was the point of returning dykelands to their drained (second-nature) state if farmers would be left to apply the same, primitive practices to maintain them?

The term was used in both of its senses in 1939 by H.R. Hare in a study of the Tantramar marshlands of southeastern New Brunswick. After reviewing both the deterioration of the land and the varied practices of the farmers, Hare paid special attention to "the personnel of the residents of this district," encouraging further study to "develop a rehabilitation program for these farmers," phrasing the matter so that it applied to both the restoration of the marshes and the re-education of the owners. For his part, Archibald, writing about the dykelands in 1944 in the *Canadian Geographical Journal,* seemed prepared to have the current marsh owners replaced by individuals returning from war, observing how "the rehabilitation of returned men and of agriculture, after the war is over, would seem to be closely associated problems."[124]

In this context, Archibald led the charge for federal action. He managed to convince his minister, J.L. Gardiner, to support an investment of $200,000, which appeared modest compared to the $6 million already invested in the PFRA by the early 1940s. Nevertheless, on three separate occasions in 1942 and 1943, the Treasury Board rejected these requests. Archibald had a hard time understanding how a project grounded in "modern engineering" could be turned down. For his part, Gardiner explained how his colleagues viewed the matter as a provincial concern, and more specifically a matter for the marsh owners who were responsible "for maintenance for the marsh lands, dykes, and drainage which is charged up against the land itself ... I shall be pleased to have the matter further discussed but I am afraid that the conclusion arrived at will be the same as that reached in the past."[125] The best that Gardiner could do was to secure $10,000 "for the purpose of making a survey of the marshland problem in the Maritime Provinces," as if it had not yet been sufficiently studied.[126]

This was a hard pill for the marshland experts to swallow. The New Brunswick and Nova Scotia legislatures passed motions in early 1943, noting that Ottawa had already "recognized like obligations in connection with reclamation projects in Western Canada and elsewhere in the Dominion." All that the Nova Scotia House of Assembly asked was that funds be appropriated "for the opening up of the main waterways and for the construction of the foundation of the dyking system, upon condition that private owners will take care of the lateral [smaller] ditching and the secondary dyking necessary to complete the work."[127]

It soon became clear, however, that one of the major obstacles to securing federal funds was precisely the suspicion that the farmers could not be trusted to hold up their end, a conclusion that naturally flowed from the studies that questioned the owners' practices. This suspicion was on display in April 1943 during hearings of the House of Commons Special Committee on Reconstruction and Re-establishment, which brought in numerous witnesses, including Archibald and the premiers of the two pertinent provinces, who made the case for rehabilitation. However, their arguments did not seem to convince several MPs from New Brunswick. Douglas Hazen wondered: "If we put these dykes back and allow every farmer to go on his own initiative are we going to be any better off?"[128] For his part, Douglas Hill had no confidence in returning the lands to the marsh bodies. He wanted to avoid

> letting your little local associations carry on in the haphazard way they have been doing in past years ... We must have some competent authority responsible for the maintenance of those lands once the improvements have been made, or replaced, because these individual farmers have shown by their practice during the past years that they cannot be relied upon to provide proper and adequate maintenance.[129]

Similar concerns about the farmers' reliability were apparent in hearings held across Nova Scotia a few months later in connection with the province's Royal Commission on Provincial Development and Rehabilitation.[130] Particularly revealing was an exchange between the commissioners and Roy DeWolfe, who owned land that formed part of the Bishop Beckwith Marsh Body near Kentville, where one of the hearings was held. DeWolfe was pressed by his questioners to admit that the dykeland problems were "probably due to neglect by the farmers." This admission led the chair of the commission, R. MacGregor Dawson, to observe: "We have no assurance that if [the land] was reclaimed the same thing might not happen again."[131]

But the clearest indication of how low the farmers' stock had fallen came only a few weeks after the Royal Commission wrapped up its hearings in Amherst. In early September 1943, most of the experts who had testified in one way or another over the previous years reassembled in the same Nova Scotia town for what amounted to a summit meeting regarding the future of the marshlands. This was the meeting described at the beginning of this chapter that had brought Archibald to the region and enabled him to take a tour that brought home the severity of the crisis. At the same time that the experts were meeting, storms and high tides were creating a situation that Archibald described as "very urgent." In that context, he regretted that "there is not any money available to assist the marsh owners in getting drag lines where ever they may be available in the Maritime Provinces with experienced operators to immediately patch the gaps in these dykes."[132]

The meeting, chaired by Archibald, brought together roughly fifty men. There were delegations from the federal, New Brunswick, and Nova Scotia governments that included both high-level bureaucrats and agricultural engineers as well as individuals connected with the Nova Scotia Agricultural College in Truro, the experimental farms in the region, and the PFRA.[133] In addition, Archibald noted the presence of "a strong delegation of practical marsh owners," the modifier underscoring the fact that their knowledge came from experience, and not technical training. Archibald conceded that the marsh owners "made some of the strongest contributions" at an opening session. They "considered a broader approach to marsh land engineering problems and some had some small experience in mechanizing of dyke building and ditching."[134] Nevertheless, this session also included repeated condescending reference to farmers who practiced tiding, and ended with the experts and the "practical marsh owners" going off to continue their conversations separately.

With the farmers in another room, the professionals and bureaucrats were tasked with coming up with a framework for emergency dyke repair, the system that continued until creation of the MMRA.[135] The repairs would be watched over by what came to be known as the Maritime Dykeland Rehabilitation Committee, which would consist of four engineers, four marsh owners, a soil specialist, and an official from the federal Department of Agriculture. There was some appropriate recognition that specific marsh owner concerns might not be adequately handled by this committee, and so it was decided that there should also be "regional committees" of owners who would be consulted "when considered necessary." However, it was clear that such consultation was not central to the process.[136]

More substantively, the Dykeland Rehabilitation Committee was responsible for devising a policy for the rehabilitation of marshlands so that costs would be divided between Ottawa, the provincial governments, and the private property owners. The committee would collect detailed information about each marsh body and "report on which are worthy of reclamation." It was also supposed to look into the means for introducing heavy machinery and for drafting new legislation to govern the marshes, the latter suggesting that the local autonomy that had been at the heart of the existing legislation might soon be challenged.[137]

As for the farmers, whose interests were central to the experts' discussions, the minutes of their separate meeting were short on details. In the only concrete proposal that was recorded, the marsh owners called for creation of a committee "consisting of one member from each county interested ... to make recommendations as to the needs of rehabilitation."[138] When the two groups reassembled to close the meeting, all of the proposals from the two groups were accepted, although it remained unclear if, or how, the proposed marsh owners' committee would exercise much influence. In any event, the paper trail provides no reference to the operations of this committee once it had been established.

WITHIN WEEKS OF THE meeting at Amherst, the Dykeland Rehabilitation Committee was up and running, even though it had not yet been provided with a budget. It operated out of the experimental farm in nearby Nappan, under the direction of W.W. Baird, the farm's superintendent since 1913. From its creation as one of the original experimental farms, the Nappan facility had had a close connection with dykeland farming. Its original property included seventy acres of drained marsh, where a variety of "problems associated with the protection, drainage, cultivation, fertility and cropping of these soils [were] studied."[139] For instance, between 1922 and 1931, the farm set aside "several marshland flats, of approximately one acre each ... to measure the response of crops to various commercial fertilizers, manure and lime"; the results indicated that improved practices could "substantially increase the average yields of crops grown on these areas."[140] During a visit to the farm in 1949, just as the MMRA was being created, the deputy minister of agriculture observed that "production on the marsh lands at Nappan have been developed to a point where, if it could be duplicated on even a portion of the marsh lands in that part of the country, the basis of successful livestock production would be established."[141]

While the experts at Nappan had studied marshland farming for decades, there is no evidence that they were involved in transferring this knowledge to the farmers – that is, until the dykes and aboiteaux had deteriorated and emergency action was required.[142] In that context, with a small budget provided by Archibald, Baird's committee hired thirty-five students from the "marshland" universities, Acadia and Mount Allison, to go out into the field to prepare an inventory of emergency work that needed to be done, in the process providing further detail about the extent of the crisis.[143] For instance, at Harvey Bank in New Brunswick, along the Shepody River (where the MMRA would build a dam in the early 1950s), there were seventeen farmers who

> depended almost entirely upon the marsh for hay. As the tide came over the marsh in July of this year practically no hay was cut ... The loss of this

marsh body means practically the loss of an agricultural community as with their main source of livestock feed gone these farmers will be forced to leave their farms and seek employment elsewhere."[144]

According to the dykeland committee, 600 feet of dyke were out at Harvey Bank, but this was "spread out over a considerable length with gaps ranging from fifteen to seventy-five feet ... In many cases, these dykes are right on the shore and they lack the foreshore which is so necessary if dykes are to withstand the rush of the tide." To deal with the crisis, at least temporarily, Baird's committee recommended "the construction of approximately 1,700 feet of new dyke which will cut off the exposed sections and repairs to another 215 feet where exposure is not so severe."[145]

The cost for this particular project, one of twenty-four that Baird identified for immediate attention, was $4,000, and the total for all of the projects was roughly $32,000, but to cover those costs the federal government needed to appropriate funds. Even by 1943 standards and in the midst of a war, this was not a large sum, the expenditure for the PFRA (to take one pertinent example) reaching nearly $2 million in 1942–43 alone.[146] Nevertheless, Ottawa's support could not be taken for granted given the earlier unwillingness of the Treasury Board to approve recommendations from the minister of agriculture. In 1942, J.L. Ilsley, the minister of finance, who was also an MP from the region, had blocked efforts to get federal funds for dyke repair, viewing them as impossible to "undertake in time of war."[147]

Indeed, throughout late 1943 and into 1944, Ilsley refused to commit any federal support, in spite of persistent lobbying by Archibald, who explained to him how "every day this construction work is delayed the loss becomes more serious through increased deposits of silt on productive land, the washing out to sea of valuable dyke lands, and the further destruction of dykes and aboiteaux on adjacent marsh bodies."[148] Archibald and his colleagues tried to convince the minister that the situation could be considered a "war emergency," because the flooding of the marshes stood to "jeopardize the highways and railroads" causing "a serious setback to the movement of our troops and war material."[149] But even this did not do the job.

The logjam was broken only in the spring of 1944, as crops for that season were already in jeopardy, when the two provinces agreed that they would put up funds that would be matched by Ottawa. The original proposal along these lines, supported by the minister of agriculture, but not by Ilsley, would have created a fund of $400,000.[150] In the final plan, however, the federal contribution was set at one-third of a fund totalling only $50,000, the other two-thirds paid in equal parts by the provinces and the marsh owners, most of whom would have been hard-pressed to participate given their straitened circumstances. Even with this drastic reduction in Ottawa's financial exposure, the federal government insisted that this was a one-off justified by "the extraordinary circumstances of war emergency" and authorized by the War Measures Act.[151] As Ilsley explained to the Nova Scotia minister of agriculture: "It was with some reluctance that I agreed to [the 1944] expenditures and there is even less reason for following the procedure for another year."[152]

For skeptics such as Ilsley, the evidence from the 1944 construction season was mixed at best. The Dykeland Rehabilitation Committee reported that it repaired over 25,000 feet of dykes and aboiteaux, in the process keeping 5,800 acres in production. For instance, in the case of the Dentiballis Marsh in Nova Scotia's Annapolis Valley, where the dyke had given way in 1943, resulting in the loss of most of the hay crop, the committee helped fund the repair of 300 feet of dykes that brought 283 acres back into production.[153] Oddly, however, the committee's own data showed that the entire area at Dentiballis would "probably [have been] protected by farmers unaided." For all of the Nova Scotia projects, the committee reported that nearly 90 percent of the "reclaimed" land could have been protected by the farmers without any government aid.[154]

These data had relatively little basis in fact, because as J.W. Byers, formerly of the Nova Scotia Marsh Survey and now serving on the committee, observed: "It is difficult to estimate what work would have been done if government aid had not been available."[155] Nevertheless, the information underscored the long-held sense that the farmers were somehow trying to game the system and were taking government funds when they could have been using their own, which flies in the face of the facts on the ground. Archibald wanted the farmers to put up one-half of the costs for repairs (as opposed to the one-third in 1944) were the program to continue, and even then only to "meet real emergencies ... I think that the arrangement for this year [1944] was decidedly over-generous to the farmers."[156] For his part, Ilsley viewed the arrangement as having encouraged "the dyke owners to refrain from doing any work at all on their dykes until they are assured that government will undertake to bear two-thirds of the cost."[157] Even Grand-Pré, usually viewed as the model marsh body, was the object of suspicion because it went ahead and did work on its own, without government approval, according to Byers, "probably hoping that the rumour of government aid would be true."[158] For his part, J.A. Roberts, who sat on the Dykeland Rehabilitation Committee, asserted that the farmers were lazy, claiming in an interview that "half the battle was getting them to show up at the same time."[159]

This cynicism stands in contrast with the perspective that W.W. Baird developed from his direct experience with the deterioration of the dykes and aboiteaux as chair of the dykeland committee. Because he doubled as superintendent of the experimental farm at Nappan, Baird's hands-on involvement with the first year of the emergency repair program was chronicled in the daily journal that documented all activities at the farm, including the comings and goings of its staff. Starting with the September 1943 meeting that led to creation of the Dykeland Rehabilitation Committee, the journal frequently noted that "Mr. Baird was away." In early May, he met with the provincial ministers of agriculture at Amherst, from where they boarded the train to travel to Ottawa on "marsh business."[160] Baird was also out in the field, most notably at Harvey Bank in New Brunswick, where he travelled on numerous occasions, his committee having warned that "the dykes might go with the next high tide."[161]

Having seen the situation on the ground, Baird not only defended the work carried out in 1944 but imagined an even more ambitious program for 1945. He proposed that Ottawa contribute $300,000, arguing that "at least 15% to 20% of our dykes and aboiteaux are ready to go at any minute, should a bad storm and high tides come together." In addition, there was the matter of projects that had been started in the program's first year but not completed.[162] Byers explained at the end of 1944 that in Nova Scotia "about 800 acres of dyke land is out to tide because the work could not be completed, or because it was not considered economical." He observed that at Belleisle "about half the job was done, but could not be finished"; at Kennetcook "the work was almost completed and then lost through an extra high tide"; and at Card's Beach "the ground was too soft for a bulldozer and it was decided that the project was not economical at present."[163] The situation was even worse in New Brunswick, where there was less dykeland to deal with than in Nova Scotia, but a larger area of land (1,000 acres) was still "out to tide because the work could not be completed."[164]

Baird's expanded program was a non-starter for his superiors in Ottawa. Archibald wrote to his deputy minister early in 1945: "Mr. Baird's estimates are entirely beyond reason, and I am sure that Mr. Ilsley will not consider them. I am, however, trying to get Mr. Ilsley to consider a small amount equivalent to what is provided in this year's special war appropriation."[165] For his part, Ilsley, in spite of his serious reservations about helping feckless farmers, recognized that "both the Dominion and the Nova Scotia governments might be open to serious criticism if the money already spent were to be lost because of failure to bring the work to completion."[166] In the end, Ottawa found it difficult to get out of the business of supporting the repair of the dykes and aboiteaux once it had started; and so it paid gradually increasing annual contributions, to be matched by the provinces and the marsh owners, each year hoping that it would be the last before a more permanent system was put in place.[167]

Baird was proud that between 1944 and 1947 his dykeland committee had managed to complete 130 projects, resulting in the construction of over twenty-seven miles of dykes and aboiteaux, but in spite of this success almost everyone connected with the yearly appropriations for emergency repairs, what the federal minister of agriculture called "hit and miss methods," recognized that a more lasting solution was required.[168] These annual appropriations were never designed to replace the weakened protective structures, which continued to require repair, new breaks emerging even as the committee was fixing older ones. For instance, Archibald expressed unhappiness in July 1945 that funds were not immediately available to solve an emergency situation connected with a failing aboiteau along the Canard River in Nova Scotia:

> It is a pity that the money is not available now ... so that the work might be proceeded with during the month of August which is a good time for construction in so far as tides and weather are concerned. If this dyke does go out it will probably cost at least two or three times to again restore it [than if] the new sluice were put in immediately.[169]

Writing along the same lines, Baird noted "that the cost of repairs increases to more than double if repairs are not made promptly after breaks."[170]

A further complaint about the system of emergency repairs came from Charles Logan, one of the few marsh owners whose voice emerges from the archival record. Logan, a farmer from Amherst Point, where he served for sixty years as an official of the marsh body, was involved in all of the discussions leading to the creation of the dykeland committee, on which he was now a member. In June 1946, he reported to A.W. Mackenzie, the Nova Scotia minister of agriculture, about a

> bad break in what is known as the McGowan Body of Marsh at Amherst Point. About 500 acres of land is inundated and a serious loss of crop will result. A small aboiteau was carried out, and usually when the marshes are covered at this time of year it damages our crop for two years at least; and

our dykes on this particular body are in a bad state, and due to the scarcity of labour ... men who can do this kind of work are simply impossible to get at any price.

Since a new appropriation of funds had not yet been approved by Ottawa, Logan was at his wits' end, fearing that unless some arrangement was arrived at soon, "we will lose our crop entirely and losses will run into thousands of dollars."[171]

The marsh owner received little satisfaction from Mackenzie, who bemoaned the fact that he could only pass the news on to his counterpart in Ottawa. He observed that Logan's letter was "quite typical of what I have been receiving for the past year or more. Such breaks as these are bound to occur until we can do a complete repair job on those outer dykes," but such repairs were outside the scope of the ad hoc arrangements. Mackenzie threw up his hands, advising Logan that for the moment they were stuck with the "temporary ... three-way scheme."[172]

Undeterred, Logan then turned to Ilsley a few weeks later, pointing out to the finance minister that "we were promised at the time of the organization [of the Dykeland Rehabilitation Committee] that our [permanent] marsh reclamation scheme would be sponsored by the government as a Post War measure. Three years have passed ... and the dykes have been badly battered." He urged "that a decision be made at the earliest possible date," but in 1947 Logan was still waiting.[173] This time he wrote to Archibald, explaining how "our marsh owners are getting discouraged waiting for results." Logan pointed out that "owners of marsh at the head of the Bay of Fundy are the only people in Canada that are cultivating land below sea level ... In Holland under the same conditions the State maintains both the dykes and drains (aboiteaux). Our government should give us the same consideration." Recognizing that Archibald had been Ottawa's point person for the dykeland dossier, Logan closed: "I am appealing to you as you were entrusted with our organization three years ago."[174]

In fact, Archibald claimed in 1945 (unknown to Logan) that he had already prepared a draft of a "Maritime Rehabilitation Act," but it was being held up until "the provincial authorities come to grips with the situation."[175] Over the following three years, until legislation was finally passed creating the MMRA, Archibald saw the provinces as the problem, engaging in further finger pointing while drained marshland remained out to sea. If the federal government were to take on the responsibility for a permanent repair of the dykes and aboiteaux, Archibald wanted to be sure that the marsh owners, already guilty of negligence in his eyes, would be held accountable by the provinces for the maintenance of drainage on their lands. This was a matter that fell under provincial jurisdiction, and so, as he put it in 1946, he kept "hammering [the provinces] to review the two provincial Dykeland Acts and bring them up to date, but as far as I know, nothing has been done. The Provincial Dykeland Acts are about as antiquated and useless as they can be."[176]

Growing increasingly negative about the situation, Archibald took the marsh owners to task once again eighteen months later, noting how they needed to have "a definite sense of responsibility in the proper ditching, the renewal of the fertility of the marshlands through the use of commercial fertilizers." He went on to describe how he had spent the summer of 1947 visiting the region but came away disappointed: "I could see no extensive improvement over previous years. The lack of drainage on those marshlands, which are still adequately protected by dykes and sluices, is an evidence of neglect, and this is certainly evidenced by the low yields." Although Archibald had been dealing with farmers such as Charles Logan for years, he now tarred them all with the same brush, and saw their redemption only through their organization into effective marsh bodies by means of new provincial legislation, "designed to meet modern needs."[177] Archibald and his colleagues wanted an entirely new structure that would force farmers to commit their resources to the upkeep of their lands, as if a lack of will had been the problem in the past. In the process, the centuries-old tradition of the marsh owners' control of their lands and the structures that protected them was about to come to an end.

Maritime Marshland Rehabilitation Administration

The legislation leading to creation of the MMRA was finally presented to Parliament in the spring of 1948. In providing this name for the agency that would be responsible for the marsh landscape for the next twenty years, Ottawa turned its back on the terminology that had been used for the temporary repairs. By tagging that committee as one looking after "dykeland rehabilitation," it was made clear that the object of attention was the landscape created by the dykes, leaving no question that this was second nature. When that name was replaced by a focus on rehabilitating "marshland," it almost sounded as if the government were going to return the marshes to their pre-European form, when in fact the goal was to ensure that the drained landscape would remain intact. Calling this landscape "marshland" was to naturalize it, conflating first and second nature. Indeed, J.S. Parker, the founding director of the MMRA, recognized the difference in terms when he made reference to "the marshlands, or dykelands as perhaps they should be called."[178]

More substantively, the act creating the MMRA provided the framework for "the construction and reconstruction of dykes, aboiteaux, and breakwaters." Ottawa was prepared to put up over $3 million, although the bill would run to more than $30 million by the time the MMRA wound up its operations in 1970, at which time it handed off the responsibility to the provinces, as laid out in the act. However, this federal support was conditional on the passage of provincial legislation committing the provinces to oversee the "reconditioning and construction" of the system of drainage ditches on the inland side of the dykes, "either with or without the assistance of the marshland owners."[179] At the very least, Ottawa wanted to be sure that it would not be stuck having to force the irresponsible farmers to shoulder their share of the burden, a concern that came out loud and clear when the House of Commons debated the bill.

This debate began inauspiciously when the MPs had difficulty focusing on the subject at hand. Following introduction of the bill, the House became a committee of the whole and immediately turned its attention to flooding in British Columbia, soon moving on to comparable problems in Ontario. After lengthy interjections along these lines, the committee chair remarked: "I must call attention to the fact that what we have before us is a resolution to assist the provinces of Nova Scotia, New Brunswick, and Prince Edward Island."[180] Discussion then returned to British Columbia, before the merits of the MMRA legislation were finally addressed.

In presenting the legislation to the House, the minister of agriculture expressed Ottawa's willingness to take the lead, claiming that the dykes would never have existed without a strong central authority. As Gardiner put it: "If it were to be done at all it had to be done on the same basis as the French did it in the first place and as the British did later on; it had to be directed and partly financed by some central authority and put into shape where the local people could handle it."[181] This was a gross distortion of the record, which was marked, as we have seen, by an absence of central control and by the dominant role assumed by local marsh owners. But that narrative did not suit the mood of the time, in which both levels of government viewed the farmers as part of the problem and centralized control as the solution.

In that context, Douglas Hazen, an MP from New Brunswick who had been involved with the marshland dossier during the 1940s, reminded the House about the farmers' shortcomings:

> For some years past the owners of the land failed to keep up these dykes; they failed to look after their own property, and as a result the dykes went down ... Now if public moneys are to be voted and spent on erecting these dykes and putting these lands in shape for their owners, I think it is essential that arrangements be made to see that the owners of the land keep up the dykes after they have been built.[182]

These arrangements were laid out in the largely identical pieces of legislation passed in Fredericton and Halifax in 1949, most of whose provisions pertained to a new system of marsh bodies.[183] While such organizations had been operating under provincial legislation that was first introduced in the 1760s, no federal or provincial support for rebuilding the dykes and aboiteaux would be provided to a marsh body unless it committed itself to an entirely new set of rules, on the face of it starting from square one. Marsh owners, even in a situation such as at Grand-Pré, where the marsh body had existed since the Acadian deportation, needed to petition the provincial government stating that they were prepared to accept the new regime. The petition needed to have the support of two-thirds of the marsh owners, holding the majority of the land within the area to be protected. Once the petition was accepted by the province's newly created Marshland Reclamation Commission, the marsh body would no longer be subject to the older rules and regulations, which people such as Archibald had found to have been at least partly responsible for the failure of the owners to keep up their lands.

In keeping with that sentiment, the new rules significantly reduced the range of activities to be watched over by the marsh bodies. For instance, in Nova Scotia the previous legislation from 1900 described at great length how bodies could protect their lands from the tides through construction of "any dyke, aboiteau, weir, dam, [or] breakwater"; could route water off their lands and toward the aboiteaux by way of "any drain, ditch or watercourse"; and could engage in tiding, by allowing entry of "water for the manurance, building up, and improvement" of the dykelands.[184]

By contrast, the new provincial legislation did not specify that the marsh bodies were solely responsible for any of the works that had for centuries been their concern. Instead, the province was empowered, on its own or in partnership with another body, "to construct, recondition, repair, maintain, conduct or operate works" pertinent to dykeland farming. In practice, this meant that the province would work with the MMRA to reconstruct the dykes, and with the marsh owners to look after the drainage ditches. As for tiding, it was not even mentioned in the 1949 provincial legislation, which was perhaps to be expected as the marsh owners' knowledge was being marginalized.[185]

While the activities for which the new marsh bodies were responsible were reduced, the mechanisms for managing their resources were increased, a change entirely in keeping with the often-expressed view that marsh owners had long been cavalier in applying their resources to the job at hand. As in earlier versions of the Marsh Act, the new legislation detailed how property would be evaluated, rates would be set, and fees would be collected; however, the legislation also mandated marsh bodies to establish reserve funds. Archibald and his colleagues were convinced that the old marsh bodies had been incapable of dealing with emergencies because they had failed to save, literally and figuratively, for a rainy day. In 1947, he criticized the Nova Scotia marsh legislation that "practically prohibited any marsh body from setting up a reserve to meet any emergencies in repairs and replacements."[186] In order that the marsh owners become ants instead of grasshoppers, the new legislation required that 1 percent of the assessed value of all dykeland be placed annually in a reserve fund, continuing until the fund equalled half the value of the property in the marsh body.

In the case of the Grand Pre Marsh Body, with property assessed at slightly over $44,000 in 1950, at least $440 would need to be set aside annually until the reserve reached $22,000, no small amount for a marsh body of the time, or, to put it another way, more than the entire federal appropriation for emergency repairs across the region at the start of that operation.[187] As Austin Taylor, the New Brunswick minister of agriculture, noted in terms of his own province's legislation, marsh owners would not on their own "provide for taxation until damage was done; therefore the reserve fund would take care of this. In a depression, many marshowners would leave repair work undone as they could not pay the bills. By setting up the emergency fund, this problem would be met."[188] There was a certain internal logic to Taylor's argument if farmers could actually find the necessary funds – but that was a very large *if*.

In the end, the marsh owners had little choice but to accept the new rules of the game, which offered them freedom from sole responsibility for the construction, repair, and upkeep of their dykes, aboiteaux, and drainage ditches in return for closer scrutiny over their now much more limited affairs. As a result, every set of marsh body minute books that I could find indicated substantial, and sometimes unanimous, support for their reconstitution, which in certain ways made little difference in terms of day-to-day affairs. For instance, the individuals who directed the Dentiballis Marsh Body under the old regime were the same as those who did so under the new one, the only change being a renaming of the board of commissioners as the executive committee, both of which were elected annually by the proprietors.[189] The seamlessness of the transition was similarly evident at Grand-Pré, whose proprietors met on several occasions over the course of 1949 and early 1950 to weigh the consequences of the new regime. They met for the last time as the "Grand Pre Dyke" in March 1950. In its last bit of business, the clerk announced that all outstanding rates needed to be paid immediately because debts "are not to be carried over to the new setup." With that, the meeting adjourned, and immediately "a meeting of the Grand Pre Marsh Body opened."[190]

In spite of this appearance of continuity, a profound change had taken place, ending, in the case of Grand-Pré, two centuries of local autonomy. Even before the proprietors met for the last time under the old regime, the new executive committee received a letter from J.S. Parker, the director of the MMRA, "requesting this committee to appoint patrolmen for the running dyke," an unprecedented intrusion on their ability to look after their own affairs.[191] Parker's intervention led executive committee chair L.H. Curry to ask him on several occasions: "Has the commission taken over the Grand Pre Marsh Body?"[192] Parker set Curry straight by commenting that "the Dominion does not actually take over the Marsh Body." He outlined the financial support that would be forthcoming for reconstructing the dykes, and assured Curry that he "did not like to think that the Body or the dykes are to be turned over to the Dominion. That is just not the case."[193] Technically, Parker may have been right, but in practice the marsh owners were no longer in charge. That mantle now belonged to the MMRA, whose experts reshaped the marsh landscape, for better or worse, over the next twenty years.

2
Reconstruction

Between 1949 and 1969, the Advisory Committee of the Maritime Marshland Rehabilitation Administration (MMRA) met thirty-eight times, but on only one of those occasions was it joined by marsh owners, whose lands were at the heart of its activities. Mandated by the federal legislation establishing the agency, the committee was responsible for approving projects that had been proposed by the marsh bodies, supported by the provincial governments, and vetted by the agency's professional staff at its head office in Amherst, Nova Scotia. No funds could be disbursed without the approval of the eight-member committee, only one of whom held his position as a representative of farmers, the other seven sitting on the committee as delegates of the federal or provincial governments. But this was an improvement on an earlier iteration of the Advisory Committee that had no explicit representation for farmers, while including a member representing the two national railways, whose lines across New Brunswick and Nova Scotia also required protection from the tides.[1]

During its twenty years of deliberations, the Advisory Committee considered hundreds of projects, ranging from the fairly straightforward reconstruction of existing dykes and aboiteaux, to the elaboration of plans to construct large tidal dams across a number of the region's major rivers. In order to carry out its mandate, it frequently invited visitors to attend its meetings, usually members of the MMRA professional staff or representatives of the provincial marshland commissions, which had been established to channel projects from the marsh bodies to Amherst. However, at its fourteenth meeting in September 1952, held in Wolfville, Nova

Scotia, as part of an annual fall tour of the dykelands, it was also joined – for the first and only time – by several marsh owners who came to plead their case.

These visitors, from the Castle Frederick and Falmouth Great Dyke Marsh Bodies, took advantage of the committee's presence in the vicinity of their properties along the Avon River to argue for construction of protective structures that would allow sections of their land that had gone "out to sea" to be brought back into cultivation. At the previous meeting, the Advisory Committee had ruled against such action, weighing – as was their norm – whether the costs involved were in line with the probable increase in agricultural production. While the committee was willing to invest $300 per acre to protect parts of these marsh bodies by reconstructing existing dykes and aboiteaux, it refused to provide $700 per acre to start from scratch where the structures had washed away. The committee was effectively restating a refusal of the same file in 1950, when it had observed that "the part of the marsh now out to tide is of questionable value because of the high ratio of dyke [to be built] to area [to be improved]."[2]

One after another, farmers came forward "to give their reasons for wishing tracts reclaimed, and their plans for the use of that marsh." One explained how

> all of his cleared land [within the Castle Frederick Marsh Body] suitable for agriculture was in use. At present, he has to buy hay every year ... If the additional marsh were reclaimed, he expected to have sufficient crop land to carry his stock. He expected that the crops grown on the reclaimed marsh would more than pay for the cost of maintenance. It would simply be good planning for him to keep the dykes up.

Similarly, a farmer from the neighbouring Falmouth Great Dyke Marsh Body observed that "he had about 35 acres of upland in cultivation and had no other marsh. He wished to increase his [live]stock as the apple industry was declining, and needed more land."[3]

The committee discussed the matter with the farmers, who then left the room so that a decision could be reached as to whether $30,000 should be invested to bring forty-two acres of dykeland back into production. Clearly, the marsh owners made a good impression because the committee members had nothing but praise for them. J.A. Roberts, who was also the secretary of the New Brunswick Marshland Reclamation Commission, was "satisfied that the men need the land and that they seem in earnest in their intention to use it." In the end, the committee reversed its previous decision to put off work, now that it had "personally visited the tracts under discussion," had "heard the statement of the owners of their plans for the use of the marsh," and had "felt that the owners were sincere in their intentions."[4]

This was an exceptional moment when the voices of the marsh owners could be heard in the affairs of the MMRA. Although the Advisory Committee would have other tours of the region, the proprietors were never again invited to speak for themselves. And more generally, when other opportunities arose for farmers to participate, the MMRA chose to marginalize them. For instance, in 1949, when the newly formed agency was organizing a conference with various interested parties to discuss how drained marshland should be used once it had been protected, it drew back from an initial idea to let farmers into the second day of a two-day meeting sponsored by its Marshland Utilization Committee, ultimately staging the entire affair behind closed doors.[5] As the *Amherst Daily News* put it in a headline: "Nova Scotia and New Brunswick Officials Hold Private Session in Local Offices."[6] J.S. Parker, the director of the MMRA, called this a meeting of "agricultural workers," but that apparently meant professionals who worked on agricultural issues, and not farmers.[7]

Given the frequently negative view of marsh owners over the decade leading up to creation of the MMRA, it is not entirely surprising that they were kept at a distance, particularly as the agency built up a professional staff at its head office in Amherst. On the face of it, individuals with expert knowledge were marginalizing those with more local forms of knowledge, but the situation was not that simple. As the MMRA went about rebuilding the dykes and aboiteaux, restoring second nature in the process, the marsh owners were rarely consulted, but their centuries-old techniques were integrated into the engineer-led projects that now included the use of heavy machinery and the prefabrication of timber for the aboiteaux. This situation was on display in the film *Les aboiteaux,* as the skills of the fictional Placide were required to install the new aboiteau. In a still from the film shown in Figure 2.1, we see the real-life Ernie Partridge supervising a crew that was adhering to new construction methods introduced by the MMRA, building on practices with which Ernie was familiar as on old-school dyke builder.

This merging of old and new was trumpeted in the MMRA's first annual report: "In the past, protective structures had been constructed mainly on the basis of practical experience but without the aid of sound design and planning. In considering new designs the advice of those having practical

FIGURE 2.1 Rebuilding an aboiteau. Les aboiteaux, *copyright 1955 National Film Board of Canada. All rights reserved. Collection: Cinémathèque québécoise.*

experience was considered and very often applied." Similarly, Robert Peterson, a soil mechanics engineer with the Prairie Farm Rehabilitation Administration (PFRA), praised the MMRA during a visit to the region in 1951: "In the case of the aboiteaux it is felt that the MMRA Engineers have very wisely selected the best procedures used by the early dyke builders but have greatly facilitated the work by use of modern techniques." As for the dykes, Peterson observed how "some persons apparently took the view that it was a matter of discarding the old techniques and using large earth moving equipment to carry out the job quickly and efficiently." However, the MMRA, much to its credit, realized that "it is impossible to use heavy earth moving equipment in soil conditions such as are found in this area," leading it to maintain "the present type of construction."[8]

The combination of practical experience and scientific technique was at the heart of what the MMRA called "reconstruction," the rebuilding or replacing of the dykes and aboiteaux that had been protecting marshlands from the tides for centuries. By the early 1960s, however, this process, which provides the focus for this chapter, was largely complete, as 236 miles of dykes and 402 aboiteaux had been either repaired or rebuilt, protecting roughly 50,000 acres of dykeland that formed part of farms totalling nearly 300,000 acres, once the uplands were also taken into account.[9] There was a pronounced slowing down of reconstruction activity by the late 1950s, reflected in the Advisory Committee's less frequent meetings; and when the committee did meet, it sometimes rejected more proposals than it recommended, something that had not previously been the case. There was a sense within the committee that weak projects were "being brought forward now, because it is assumed that this is the last opportunity to have them considered for reclamation."[10]

Included among those weak projects were proposals from marsh bodies that had also been turned down on previous occasions. Indeed, the MMRA decided not to rebuild structures that might have protected a further 14,000 acres across the region, a significant part of which was located in Nova Scotia along the Cumberland Basin, the northeasternmost extension of the Bay of Fundy. For instance, the Advisory Committee rejected numerous requests for reconstruction work from the John Lusby Marsh Body, where over 1,200 acres were never brought back from being "out to sea"; the land, having reverted to salt marsh, ultimately became part of a National Wildlife Area.[11]

With its reconstruction work nearing completion, the committee did not meet at all between September 1960 and July 1964 since there was nothing for it to approve; and it met just five times after the summer of 1964, rarely discussing dykes and aboiteaux. Instead, talk turned to finding a formula for the return of the protective structures to the provinces, which had been the endgame from the start. And while the MMRA would continue to operate until 1970, most of its activities and nearly all of its budget during its last decade pertained to the construction of a series of large tidal dams across a number of the region's major rivers, protecting a further 32,000 acres of dykeland, in the process creating a third nature, which provides the focus for the second part of this book.

During the 1950s, however, the restoration of the old protective structures dominated the affairs of the MMRA, a process that worked against a simple binary dividing expert and local knowledge. As we will see, Amherst became the hub for a self-conscious community of specialists – engineers, draughtsmen, agricultural economists, and soil experts – who became internationally recognized leaders in their fields. At the same time, the professionals never entirely rejected the techniques of those who had been working the land long before the intrusion of the state, and the engineers became guardians of "local" knowledge in their own right, as their expertise was literally grounded in this particular landscape.

By adapting their expert knowledge to the particular marshland environment, the MMRA personnel displayed what James Scott meant by "mētis," a term that he applies mostly to individuals without technical training, since it refers to "a rudimentary kind of knowledge that can be acquired only by practice and that all but defies being communicated in written or oral form apart from actual practice." Ultimately, however, the crucial thing about mētis was that it required application of general rules "to broadly similar but never precisely identical situations requiring a quick and practiced adaptation."[12] The MMRA's engineers adapted their practices to incorporate the traditional knowledge of the marsh owners, at the same time marginalizing the farmers from decision-making. Ironically, while the marsh owners' practices were worth appropriating, the farmers were left to play a secondary role in shaping their own landscape.

Amherst

In the late nineteenth and early twentieth centuries, Amherst, Nova Scotia, was a thriving industrial centre. Between 1901 and 1906 alone, the value of goods produced in the town's factories quadrupled and its population doubled, reaching nearly 10,000. As one of the most important manufacturing centres in Atlantic Canada, it became known as "Busy Amherst."[13] One of the keys to the town's growth was its location on the main rail line connecting the region with central Canada, and one of its major factories was Rhodes-Curry and Company, a locally owned firm that produced rail cars, employing 2,000 workers in Amherst in 1905. In 1909, as part of the larger process of concentration that contributed to the deindustrialization of Atlantic Canada, Rhodes-Curry was merged with two Montreal firms to create Canadian Car and Foundry (see Figure 2.2), but during the economic downturn following the First World War, Canadian Car began

FIGURE 2.2 Canada Car and Foundry building, Amherst, NS, 1931. The small objects on the land in the background are hay barns on drained marshland. *Nova Scotia Archives, Richard McCully Aerial Photograph Collection.*

closing down its various operations in Nova Scotia. As Nolan Reilly has explained: "While Canadian Car & Foundry modernized its Montreal facilities and constructed a new plant in Fort William [Ontario], it retreated from car building in Amherst," ultimately closing the plant altogether during the 1920s, a development reflected in the decline of the town's population by 25 percent during the decade.[14]

After some short-lived activity during the Second World War when the factory assembled the Anson, a training plane for pilots, the sprawling Canadian Car complex was shuttered until part of it became the headquarters of the MMRA in the late 1940s. On the face of it, there was something incongruous about a section of this industrial property being converted to support the restoration of the dykelands, but this was the start of a new era in the three-centuries-long history of marsh farming, which was now to be modernized. A weekly newspaper for farmers cut to the chase in the headline for an article about the Amherst operation: "Marshland Reclamation Goes Scientific." The article chronicled how, in early 1950, less than a year after opening, the offices already had "more than twenty full-time employees: engineers, draughtsmen, stenographers and others, and will likely have more than thirty office men working toward reclamation of some 90,000 acres of marshland before another year rolls around."[15]

These employees were needed because "a huge volume of office work" had to be completed "before workmen begin to salvage inundated marshes."[16] The marsh bodies that operated before the MMRA came into existence did not produce the sort of detailed records that engineers now required, and so in 1949, during the agency's first summer in operation, thirty surveys were completed across the region with the participation of the federal and provincial governments as well as "the marshland owners concerned." J.S. Parker, the newly appointed director of the MMRA, observed that "it was not the intention of anyone to override the marsh owners." Advancing the spirit of mētis, he observed that "reclamation of these potentially valuable farm lands was a matter of cooperation."[17]

The information collected enabled production of "maps illustrating property boundaries, lines for proposed new dykes and aboiteaux, [as well as] present dykes and aboiteaux." Still other information from those surveys was passed on to draughtsmen and engineers who translated "figures, measurements, and other data into charts which will be used when the green light is given for construction."[18] This data was crucial, for instance, in the prefabrication at Amherst of the lumber that would be used for construction of sluices for aboiteaux (see Figure 2.3). As an instructional manual for MMRA employees noted in 1949:

> The Office Engineer will have delivered to the site of the job all sluices complete for the size and length required. The timber sluice will be cut and bored ready to set up as shown on sketches ... Hardware will also be supplied. Besides bolting together and setting up, the only work that needs to be done in the field will be sawing plank to required length.[19]

This photo is one of the few images that provide a glimpse of what went on inside the MMRA's facilities, but other aspects of the agency's operations at Amherst can be pieced together from documents that have survived.[20] For instance, in addition to the shop cutting lumber for the sluices, the construction process was also supported by the "soil mechanics laboratory ... on the ground floor of the Amherst offices." The study of soil samples

FIGURE 2.3 Prefabricated lumber for sluices, MMRA workshop, Amherst, NS. *Acadia University Archives, George Frail Fonds, 2015.035-FRA, 19. Reproduced with the permission of Pamela Frail.*

would "show what combination of sand, clay, and silt is available to make the best possible dyke for that area."[21] During the summer of 1950, its first in full operation, the laboratory collected over 500 soil samples from thirty-three marsh bodies "to obtain a general picture of marshland soil." By 1951, the director of the operation, L.W. McCarthy, confidently reported that "we have sufficient experience with marshland soils to detect troublesome foundation conditions." As a result, it became standard procedure to sample the soil at an MMRA worksite so that the lab could run tests that allowed "reports to be sent out to the district engineers [including] recommendations for dyke and aboiteaux locations."[22]

To complete the picture of the MMRA facilities, there was also a hydraulics lab located in a quarry just outside Amherst. Ernie Partridge worked there during the winter, when weather prevented him from inspecting or working on the dykes. He provided labour for tests designed to determine the amount of water that should flow through a field's drainage channels on its way to an aboiteau. A 1951 report from the lab explained the challenge in designing these channels, which measured more than 1 million feet in some marsh bodies. They needed to be large enough to accommodate exceptional runoffs, something likely to occur only once every five years; at the same time, if the channels were too large, at times of "low water, the vegetation would grow rapidly and reduce the velocity at high water stages ... Drainage channels do silt, and periodically the channels have to be cleaned to renew their efficiency at high stages."[23] In that context, Ernie assisted with experiments that used flowmeters, measuring "the amount of water, the gallons of water comin' out through a four-by-four sluice." Ernie placed rocks of different sizes at the end of the sluice to test the strength of the flow. Laughing about the job, he asked: "Can you picture pickin' up rocks with a pair of tweezers all winter-long?"[24]

In order to staff the professional positions in its Amherst facilities, the MMRA brought in men (there were no women as far as I could tell) from across Canada. Leading the way was the founding director, J.S. Parker, who came from Saskatchewan with a degree in agricultural engineering. Parker worked at the PFRA experimental station in Swift Current, where he was in charge of water conservation projects from 1935 to 1938, before serving during the war as a major in the Royal Canadian Engineers. At the end of the war, he "spent one year in Holland reclaiming dykes for the RCE," leading the Canadian effort and – although he couldn't have known it at the time – gaining pertinent experience for his time with the MMRA. Parker subsequently returned to Swift Current, where in 1948 he caught the eye of J.G. Taggart, a senior official in the Department of Agriculture, who was involved with setting up "some administrative machinery for the new reclamation program." Writing to his minister, Taggart explained how the choice had been narrowed down to "engineers associated with the P.F.R.A," which in turn highly recommended Parker as "a good administrator, an exceptional personality to get along with people, and one of our [up and] coming men."[25]

The men who joined Parker in Amherst came to the MMRA by various routes. Some, like the director, were professional engineers, such as J.D. Conlon, who was the agency's chief engineer, and L.W. McCarthy, who headed up the soil mechanics lab. Conlon was assisted by Ron McIntyre, who, like Parker, came to the MMRA from Saskatchewan, where he had received his engineering degree. As for George Frail, he worked for the MMRA as a summer student before graduating from Dalhousie with his civil engineering degree in 1951, which led to a full-time position in 1952. Over the following decade, he was the lead engineer for projects that resulted in the reconstruction of existing dykes and aboiteaux, while working as well on the construction of the Annapolis River Dam.[26] Similarly, John

Waugh started at the MMRA in 1950 while still an engineering student. He carried out surveying work during summers, but with his degree in civil engineering, he moved up the ranks. As he explained to me, he "ran the design office [in Amherst] where he had eight or ten draftsmen doing other work, like topo plans."[27]

Others came by less traditional routes. Take, for instance, the case of Constant (Con) Desplanque, who was recognized as a professional engineer only in 1967, long after he had begun working for the MMRA. Born in the Netherlands, Desplanque studied land reclamation and the upkeep of dykes prior to the outbreak of the Second World War, which saw him serve in the Dutch army before its defeat led him to the Dutch underground. Following time spent in Indonesia, he came to Canada, where he was working on a farm not far from Amherst, when, as he put it in his unpublished autobiography:

> I heard about the marshland rehabilitation program and got in touch with the people there. I got a job offer for five dollars per day, six days per week with the MMRA. I took the job and started working in November 1950. I started off as a surveyor. But it was soon too cold and snowy on the marsh. So I was asked if I knew something about hydraulics. I could not say "no." Thus I was given the chance to go over the reports of some consulting firms and see how they arrived at the dimensions of certain large tidal sluices. It took some doing and reading, but I was able to master the theory and start calculating the required dimensions for small tidal sluices.[28]

And then there were individuals with responsible positions at Amherst who had no professional credentials whatsoever. Part of a family that had been farming dykeland at Grand-Pré since the 1760s, R.H. Palmeter occupied a prominent place in the local marsh body, distinguishing himself in 1939 for having introduced the use of heavy machinery for dyke construction. In an interview in the early 1970s with J.W. Byers, in his own right an agricultural engineer with the MMRA who held a degree from Dalhousie and taught at the Nova Scotia Agricultural College, Palmeter remembered how it had become impossible to reconstruct dykes by hand because labour was both scarce and expensive.[29] Byers recalled Palmeter telling him that "if he couldn't get a machine Grand-Pre would go out to tide because men were just not available." With the use of a dragline, Palmeter was able to drastically reduce costs, but he still faced naysayers. As he put it: "Yes, the [costs] were low, but they [the marsh owners] had never handled that kind of work and they had no idea."[30]

Palmeter was willing to adapt long-standing practices, while the MMRA was interested in making adjustments of its own to integrate local knowledge into its projects. In that context, a decade after the dyke construction experiment, Palmeter was hired by the MMRA, leaving his position with the marsh body and leasing his land at Grand-Pré to become the agency's chief aboiteaux superintendent.[31] This was a significant job, the term "aboiteau" being used, as was frequently the case, to refer to both the dykes and the structures that allowed water to drain from a marsh, terminology that is still used today. Instructions drafted shortly after the start of the MMRA's operations noted: "The Chief Aboiteaux Superintendent is in charge of all aboiteaux work and authority must be obtained from him before any changes are made in construction methods." This point was underscored when the same document specified what was involved with "the construction of a 'Palmeter' type aboiteau," which was first designed by R.H. Palmeter and his brother Fred in building the Dead Dyke aboiteau near Grand-Pré in 1947.[32]

An MMRA report described how their sluice was unique as it was constructed from timber prepared for assembly before reaching the work site, anticipating the prefabricated material that would be prepared in Amherst. "When the sluice was completed it was the best design that had been made up to that time."[33] More generally, the Palmeter design was based on "utilizing ... easily obtainable material, reasonably priced." In the process, Acadian practices were adapted to the mid-twentieth century, embodying the essence of métis.[34]

Despite their different career paths, the MMRA professionals shared a certain grounding in the marsh landscape, regardless of where they came from, in the process complicating the meaning of local knowledge. Parker stayed in his position in Amherst before being promoted to a new one in Ottawa in 1961, but Desplanque remained in the area until his death in 2010; most of the other professionals hailed from the region and never left. These professionals were not simply passing through, and so represented what Tina Loo had in mind when she referred to "high modernist local knowledge." By integrating some of the marsh owners' techniques into their practices (and one of the marsh owners into their inner circle), Parker and his colleagues at Amherst reflected what Loo found in the writings in the late 1950s of the engineer Robert Peck, whom she describes as having "worked on many of Canada's and North America's large dam projects." Along the way, Peck came to see that "good engineering, like good farming, was about improvising and acknowledging the limits of what nature offered ... The knowledge promoted by Peck was a kind of métis."[35]

This acceptance of local limits did not prevent the MMRA professionals from associating themselves with the larger movement across North America designed to put agricultural engineering at the forefront of changes in how farms were organized and operated. As Deborah Fitzgerald explains in her book *Every Farm a Factory*, the middle decades of the twentieth century saw the emergence in the United States of "new professional groups coming out of the agricultural colleges, in particular agricultural engineers and agricultural economists. These experts ... learned the theoretical and abstract approaches of their disciplines, especially engineering, science, and economics. This was the first generation of college-educated agricultural experts in America." Based on the collection and assessment of data, "their education pointedly did not train them how to be good farmers; it trained them how to be good analysts and evaluators of farmers and their farms."[36]

In this context, the MMRA experts were involved not only with rebuilding the dykes and aboiteaux but also with building a certain MMRA brand, in the process touting their particular skills. In part, they did this by regularly participating in engineering conferences, such as those held by the American Society of Agricultural Engineers (ASAE). As Fitzgerald explains, the ASAE was created in 1907 expressly to distinguish its members from "less educated practitioners."[37] When MMRA professionals presented papers at such conferences, they frequently did so as a group, thereby drawing attention to the team that had been assembled at Amherst. For instance, the agency was represented at the ASAE's 1951 conference by Parker (the director), McCarthy (the head of the soil mechanics laboratory), and R.R. McIntyre (the assistant chief engineer), who collectively addressed their professional peers on "The Reclamation of Tidal Marshlands in the Maritime Provinces of Canada." Similarly, in 1958, McIntyre teamed up with Desplanque to discuss, before the Engineering Institute of Canada, the challenges of dealing with the buildup of silt in the tidal rivers of the Bay of Fundy; and in 1962 McIntyre and Palmeter explained the use of sod to cover dykes in a presentation to the Canadian Society of Agricultural Engineering.[38]

The self-conscious development of a professional community was even more clearly evident when the MMRA decided in 1958, on the tenth anniversary of the passage of its founding legislation, to create an internal seminar series to encourage professional development. The unsigned text describing the series explained, a bit whimsically, how

> after ten years of work there is a wealth of information in reports, survey plans, cost studies, note books, folders, scraps of paper, and in the memories

of the staff. However, at the present time, a great deal of this information would not be available to the public, and might be lost if, say, a number of the staff suddenly decided to take up tidal studies in South America, soil mechanics in Ceylon, or offer themselves as volunteers for space travel.[39]

With that in mind, staff would be assigned topics on which they had expertise, in some cases "merely collecting and editing existing data," while in others "some field work and travelling will be required." The seminars would "assist [staff] members in determining what information was available and possible sources for further data, and serve as a sort of examining board, to catch errors and omissions in the reports, and thus make them a more valuable reference in any study on marshland reclamation."[40]

The seminars largely built on such predictable topics as the various elements connected with the construction of dykes and aboiteaux, the fundamentals of soil mechanics, and the hydraulics of sluices and gates. There was, however, a final topic that stood apart from the others not only for its relevance to the MMRA's mission to rebuild the protective structures but also because it pertained to preserving a record of its work and making the agency more visible. This last seminar would consider "photography as a means of recording facts and presenting problems or methods," dealing with such topics as the best type of camera – both still and movie – to use, best practices for archiving images, and determining which subjects could best profit from photographic treatment.[41]

From the very start of the MMRA era, officials expressed interest in creating "a pictorial record of the marshland reclamation work," including production of an educational film, "which could be shown to marsh owners in particular and farmers in general, in order to familiarise them with the possible use and practises [sic] to which marshlands can be adapted."[42] Over the years that followed, photographs were a staple of the MMRA annual reports, and the various archival collections pertinent to the agency include photo albums documenting its numerous projects. Some of those photos, such as Figures 2.4 and 2.5, showing men at work, have been reproduced in this book.

By the late 1950s, however, just as the MMRA was organizing its seminar series, there was a shift from its use of photography as a tool exclusively for documentation to one that could also be used in the field. Just before the seminars were launched, C.A. Banks, whose name appears on a number of the agency's photo albums, authored instructions for the use of "photography in surveying." As he put it: "The science of Photography, applied to the science of Engineering, has a potential usage limited only by the

FIGURE 2.4 Excavating sluice bed, Comeau Marsh, July 1949. *Acadia University Archives, MMRA Fonds, 1998.001/2/1/1. Reproduced with permission of Nova Scotia Department of Agriculture.*

FIGURE 2.5 Assembling prefabricated aboiteau, Grand Pre Marsh Body, 1950. *Acadia University Archives, MMRA Fonds, 1998.001/2/1/4. Reproduced with permission of Nova Scotia Department of Agriculture.*

imagination of the user." Banks went on to show how photography, when used properly, can aid in surveying marshland. He explained how aerial photography, or photography from a raised platform, did not really do the job, suffering from what he called "a defect," as the photograph flattens the landscape and makes it impossible to see variations in the soil. To overcome this problem, Banks detailed a method in which two cameras would be used, so that photographs of a given location from two different angles could be merged into a single image that "shows a contour plot of the area ... The application of photographic methods is not only feasible and desirable but constitutes the only economical method of doing the job."[43]

The creation of a photographic unit within the MMRA's operations at Amherst underscored both the extent of the scientific operation that was built there, mostly during the 1950s, and the self-conscious promotion of the agency as one that embraced modern techniques. Reflecting the widespread view of the collective talents of the engineering staff, the New Brunswick minister of agriculture pleaded that they be kept together as the MMRA began to wind down in the mid-1960s. J. Adrien Lévesque called them "the most knowledgeable assemblage of engineering talent in the field of saltwater control of any group in North America."[44] The agency's scientific credentials were important as it engaged with the marsh bodies to fix or replace the protective structures, sometimes building on practices that the marsh owners had honed over centuries, and other times expressing exasperation with farmers who could never live up to the scientific standards cultivated at Amherst.[45]

Simplification

Just after its creation, the MMRA issued a document laying out the procedures for work to be done to reconstruct the dykes and aboiteaux. The instructions from Amherst began with a premise: "Assume that a small number of farmers own adjoining lots of marshland, near their farms on the upland. In the past, each man has built and repaired the protective structures, which keep the salt water off his land." This arrangement was "found to be not entirely satisfactory," now leading the owners "to unite into a marsh body."[46] The document conveyed the unmistakable impression that solitary marsh owners came together only after the passage of the legislation, both federal and provincial, that brought the MMRA into existence, in the process ignoring the centuries of collective maintenance of the dykes and aboiteaux. Of course, marginalizing the proprietors was consistent with the disdain with which they were treated as the structures deteriorated during the 1940s.

The procedures went on to explain that owners who wanted work done needed to "put in a petition to the Provincial Secretary that the section be incorporated into a marsh body."[47] Government officials expected this to be a complicated process, the federal deputy minister of agriculture asserting in 1949 that there had been "difficulties in the past ... to get the marsh owners to incorporate into marshland bodies."[48] No evidence was provided for this further example of condescension, which was at odds with the record of marsh bodies, whose minute books stretch back to the nineteenth century. Indeed, it is hard to imagine that marsh bodies could have organized as rapidly as they did without the farmers' long history of collective management. Motivated by the support that would become available once they were incorporated, most of the marsh bodies took that step within the first few years after the MMRA was created; the agency's own data indicated that all dykeland in the region had been incorporated by the end of the 1950s.[49]

While there were instances of individual farmers who seemed unhappy about the intrusion of the state, I did not find a single example of farmers rejecting their incorporation into the new regime. As Jim Bremner, whose family owned land in the Castle Frederick Marsh Body (one of the marsh bodies described at the beginning of this chapter as having met with the Advisory Committee), explained to me, the farmers "were damn glad that the MMRA came, because keeping up the dykes was getting hard."[50] While the New Brunswick minister of agriculture claimed that "farmer thinking at the time tended to reject any infringement on his responsibilities or his rights," there was significant enthusiasm for incorporation.[51] In this regard, there were farmers such as A.E. Black, the clerk of the Coyle Landry Marsh Body in New Brunswick, who wrote to Parker about his efforts in 1949:

> I have spent a lot of time the last year canvassing the marsh owners; and this summer [I worked] getting the petition for Certificate of Incorporation, which I did on my own spending many evenings and a few Sunday afternoons travelling sometimes six or seven miles on a bicycle, having no car to use. I was interested in the project and was anxious to carry it through.[52]

With the requisite number of marsh owners signed up, the proprietors forwarded their documentation to the province's newly created Marshland Reclamation Commission, staffed largely by men who were marsh owners themselves. If everything was in order, the marsh body was given both a name and a number, a foreshadowing of the close supervision to follow. In terms of how the marsh organizations would be identified, a senior Nova Scotia bureaucrat wrote to the province's Marshland Reclamation Commission in 1950 asking that each name fit a particular formula: "I think it would be desirable in each case to have the name of the corporate entity terminate with the words *Marsh Body* so that they would be called the Gay's River Marsh Body, Smith River Marsh Body, or the like." He saw a problem when "in the River Hebert group the petition is for the creation of the 'River Hebert Body' and in the case of the Masstown group the petition requests the name 'The Masstown Marsh.'"[53]

This standardization was taken a step further when each of the provincial marshland commissions created a numerical system to keep track of their marsh bodies, adopting a formula whereby each body was known as NS or NB, followed by a number in the order in which their petitions for incorporation were accepted, thus making, for example, the Grand Pre Marsh Body "NS 8" for all official business.[54] Such naming practices were part of what James Scott has described as a process of "simplification and legibility." In order to create order where there had been chaos, the new dykeland administration headed by the MMRA needed a "narrowing of vision [to] bring into sharp focus certain limited aspects of an otherwise far more complex and unwieldy reality." The question of naming reflected the agency's interest in dealing with "standardized units" that would more easily fit into the planning exercise at hand.[55]

This process continued more substantively as the MMRA watched over the details of dykeland management. For instance, in the early years of its dealings with the Grand Pre Marsh Body, the agency kept a close eye on everything from the payment of local labourers to carry out regular maintenance to the appointment of individuals to patrol the dykes. Regarding the latter, J.S. Parker asked "the Executive Committee [to appoint] a reliable man ... during the periods of these or other high tides." Although the marsh body had been watching over its dykes for nearly two centuries, Parker apparently doubted that a trustworthy person would be selected, going on to note: "I feel he should not only be appointed, but should be instructed to do this job, and to take whatever action is necessary in the event of an emergency."[56]

Close supervision was also front and centre as the newly constituted marsh bodies sought assistance for the reconstruction of their dykes and aboiteaux. Typically, they would indicate the work that needed to be done to the provincial Marshland Reclamation Commission, and if the request seemed reasonable a contract was executed. In October 1949, within weeks of being officially incorporated, the Grand Pre Marsh Body signed an agreement with the Nova Scotia government asking that the province approach the MMRA about carrying out an ambitious program. In return, the marsh body committed itself to acquire, through expropriation if necessary, all land that might be needed to carry out the project, and promised that it would maintain "all drainage and protective works in a condition satisfactory to the said Reclamation Commission." In addition, the marsh body agreed to "encourage the improvement of land management and cultural practices by means of educational campaigns, tours through marsh areas, and through any other means." This provision was entirely in keeping with the prevailing sense that rehabilitation pertained to improvement of both the land and the people who occupied it.[57]

Assuming that the MMRA would go ahead with the project, the province agreed to provide up to half of the costs connected with upgrading the marsh body's drainage channels, the same channels that were the focus of the MMRA's hydraulics lab at Amherst. In some projects, over 1 million feet of ditches needed to be dug, and since there was little in the way of machinery at the outset, the costs associated with drainage could be greater than those connected with the dykes and aboiteaux.[58] Halifax and Fredericton worried about being stuck with the bill if the marsh owners failed to hold up their end of the agreement, a matter that took on even greater importance as the MMRA frequently recommended eliminating aboiteaux to reduce reconstruction costs, a process that resulted in the building of longer drainage channels to reach the remaining structures.[59] In 1952, the MMRA's Special Committee on Drainage Problems expressed concern over leaving ditch construction to the farmers' initiative, fearing that "considerable time would lapse between the time of the reconstruction of the protective works and the reconstruction of proper drainage facilities."[60]

After acceptance by the province, the project made its way to Amherst, where the professional staff carried out a preliminary investigation that included not only an assessment of what work needed to be done with regard to the protective structures and drainage facilities but also a report on "land use, and the general type and quality of agriculture in the area." To acquire such information, and to enable work to proceed as quickly as possible "to prevent further loss of productive land," MMRA staff were dispatched to the marsh body to "set boundary posts locating the boundaries of the marsh," and to provide a survey indicating "the lengths of dyke requiring reconstruction, the dimensions of aboiteaux, and notes on construction." This information was sent to the drafting room, where

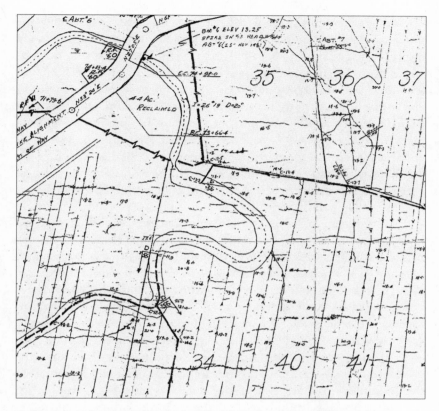

FIGURE 2.6 Bishop Beckwith Marsh Body, MMRA survey map, NS 65, 1957. *Courtesy of Eric Patterson.*

"a complete file of RCAF [aerial] photographs were available ... These photographs were enlarged and the main features of the marsh drawn on a 'Survey Sketch.' All aboiteaux were numbered for future reference in construction."[61]

Typical is this survey map (see Figure 2.6) of the Bishop Beckwith Marsh Body, which sits along the Cornwallis River in Nova Scotia, just west of Grand-Pré. Created in the mid-1950s, the map identifies both property owners and aboiteaux by numbers. For example, we can see "Abt 6," which drained into the Cornwallis, carrying water from the large ditch that runs across the left side of the map; the main ditch in turn carried water from the smaller ditches, represented here by the heavy black lines with arrows noting the direction of the flow.

Through these maps, the MMRA professionals were continuing the process of "simplification and legibility" that had begun with the naming of the marsh bodies, now inscribing on the landscape what they believed was important. Much like cadastral maps, the careful plotting of boundaries drew attention both to the existence of the new-style marsh bodies and to the holdings of individual proprietors, now numbers as opposed to the names frequently inscribed within boundaries on the few existing maps from the pre-MMRA era; instead the space was used to provide detail about changes in elevation that would be important for drainage purposes. Similarly, while earlier maps often included the word "aboideau" where such structures were located, there was now numbering to facilitate discussion regarding their disposition.[62]

These new maps reflected a certain reality but were also selective regarding what they presented. James Scott has noted how there was a tendency to make boundary lines "more geometrically regular than they were in fact, ignoring small jogs and squiggles." Moreover, such maps had little to say about what farmers were actually doing with their land, which would have only introduced "needless complication," because the maps were "designed to make the local situation legible to an outsider." Scott saw that while this exercise in "state simplification" facilitated management of this landscape, it marginalized the texture of the lives of the people who lived there, the possessors of more traditional forms of knowledge. Indeed, Scott observed that these mapping exercises suggested that "any value that the land might have for subsistence purposes or for the local ecology was bracketed as aesthetic, ritual, or sentimental values."[63]

The process of simplification was also at work as the MMRA staff tried to assess for any given marsh body how the land might be used, and what benefits it might bring if the protective structures were improved. The agency wanted to avoid investing in dykes and aboiteaux only to have the marsh owners fritter away the land through poor practices, neglecting proper drainage facilities and failing to adopt current agricultural methods. In that context, only months after taking charge, J.S. Parker returned from a tour of the region, disheartened by

> how these Bodies have gone downhill and become in a poorer condition regardless of the [emergency] assistance they have had from the Dominion and the Province. I get the impression that although the financial assistance was given, nothing was done from the long term point of view. The farmers appear to have taken the assistance for granted and in some areas there is no doubt what-so-ever that they considered it almost as a relief program.[64]

Leaning on long-established stereotypes of Atlantic Canadians as content to live off government support, Parker claimed that marsh bodies had not made needed repairs during the period of emergency relief that required them to pick up one-third of the bill. Instead, they were "setting back knowing that we would be taking over the dykes and aboiteaux, and of course now they wish to work day and night to repair them with us footing the bill."[65]

Parker viewed the situation as "almost terrible," leading him to the conclusion that "something 'shocking' will have to be done on a Land Use Program."[66] Toward that end, the Marshland Utilization Committee, the same body mentioned at the beginning of this chapter that excluded the farmers, met in early 1950. Representatives from the pertinent governments discussed how "proper utilization must go hand in hand with reclamation." Expert after expert spoke about what farmers needed to do (with the use of chemicals receiving particular attention), but without any indication of how the farmers were supposed to finance such improvements.[67] Indeed, later in the decade, J.A. Roberts, who headed the Agricultural Engineering Branch of the New Brunswick Department of Agriculture, reported that "experience indicates that costs for field drainage, plowing, liming, fertilization, cultivation and reseeding will approach sixty dollars per acre," or more than the value of most dykeland. Roberts concluded: "It is not reasonable to suppose that many farmers are in the financial position to undertake such a program."[68]

At the close of the 1950 meeting on marshland utilization, there was a proposal to create a more permanent committee to improve farmers' practices, but the idea was abruptly shelved, perhaps in recognition that without additional resources, many of the farmers would be unable to adopt new methods to manage their dykeland.[69] In the years that followed, such hectoring of the farmers continued, invariably without the commitment of support that would have enabled them to be rehabilitated along with their land. In that context, the MMRA decided that it needed to prioritize projects for farmers who had taken good care of their land in the past; farmers with limited resources, who had been unable to keep up their dykeland, would be placed at a distinct disadvantage.

Toward that end, the Advisory Committee, which would be receiving requests for reconstruction of dykes and aboiteaux, wanted a single number, "a rating that would consist of the relationship between the expected land use program and the cost per acre of the protective structures."[70] In much the same way that the MMRA wanted sharp lines on a map in order to simplify the marsh landscape and make it more legible, it also wanted a seemingly objective indicator that would incorporate a number of variables, reducing them to a single, simple score. In order to achieve this goal, the agency turned its sights to the work of yet another group of experts, agricultural economists. Deborah Fitzgerald, who drew our attention to the role of agricultural engineers in transforming farming practices, also pointed to the contribution of "agricultural economists [who] were able to articulate a more focused program that linked farm life with modernity ... The economists hammered home the notion that progress should be equated with quantity and scale more than any other measure."[71]

The MMRA was particularly influenced by the work of Gordon Haase from the agricultural economics section of the federal Department of Agriculture. Haase was the author of several internal studies regarding agricultural practices in the various marshland regions, indicating which areas had been the most productive.[72] Building on his field work, in early 1951 he presented to the MMRA's Advisory Committee a proposal for creating a cost-benefit ratio that could be applied to all marsh bodies, effectively providing the means to generate a single number to represent the "risk" of investing the agency's funds.

The committee meeting in question was attended by an unusually large assemblage of visitors, over twenty representatives of government agencies, who were on hand to hear reports from a number of specialists. Experts made presentations on such matters as testing soil samples and applying chemical treatments on dykeland, but most of the meeting focused on Haase's findings. On entering the meeting room, Haase asked twice that the details of his report be kept confidential, perhaps concerned about pushback against an exercise designed to distinguish winners from losers. He then went on to explain how it had "become customary to appraise projects of this type in terms of the costs-benefits ratio." Here was a useful tool for "comparing the effectiveness of alternative avenues of public investment."[73]

In order to arrive at this ratio, Haase needed to make a number of assumptions. He viewed the calculation of costs as fairly straightforward, since it amounted to tallying "the costs of expenditures required for structures to protect the marshland from flooding by the sea and to allow for its freshwater drainage." He insisted, however, that these costs needed to be reduced by calculating the savings from protecting, as part of the process, such facilities as railways or highways. As he explained, public agencies were "saved the costs of providing their own protection against flooding ... These benefits should be subtracted from the cost of reclamation before these are taken into account in a costs-benefit ratio."[74] As the MMRA moved into ever larger and more expensive projects such as tidal dams, the savings to other governmental agencies loomed large in providing justification for the investment.

As for the other part of the equation, Haase admitted that "the nature of the benefits from public investment in marshland reclamation is not so easily observed." Nevertheless, he proceeded to show how it was possible to calculate the contribution of an acre of dykeland in a particular region to the gross national product. His calculation was based entirely on current production on marshland that was being cultivated, so that other benefits, such as increases in production on existing marshland or the introduction of land that had been out to tide, were not taken into account. Haase concluded his presentation by arguing that the MMRA should develop a cost-benefit ratio, even though he recognized that "such a ratio may be rough or approximate ... [The MMRA needed to] conserve the best lands first, that is, lands which are best used by the owners, or where the present benefits are the highest."[75]

Having finished his presentation, Haase was questioned by members of the Advisory Committee, one of whom pointedly asked whether "the farmers had been asked their opinions as to the reason for the present farming conditions. Mr. Haase replied that no opinions had been recorded, only facts were sought."[76] Of course, Haase had recognized that his own "facts" were less than perfect, a point underscored by a senior official in Ottawa who questioned Haase's calculations and advised against "distribution of the portion of the report dealing with project evaluation."[77]

More broadly, there was criticism of the assumption that the MMRA should privilege those regions that were already highly productive. According to J.G. Taggart, the federal deputy minister of agriculture, the MMRA had been created to carry out work that "in most cases could not be done on an economic basis by the owners of the land. If the individual projects are now to be judged by strict economic standards – that is the cost of the rehabilitation in relation to the value of the particular parcel of land protected – we are going contrary to the purpose of the act." Taggart went on to observe that "one of the main purposes of reclaiming these lands is to help to strengthen the farming communities which depend in part on marshlands for hay and pasture to supplement the relatively low productivity of the 'uplands.'" Taggart regretted that such intangibles, much like the information disregarded by the MMRA's maps, were made to disappear from view "by merely dividing the number of acres directly involved into the cost of the project."[78]

In spite of such reservations, the MMRA continued to search for a formula to rank marsh bodies. In early 1954, senior officials responsible for agriculture from the three pertinent governments met in Moncton, in large part to discuss how decisions were made by the Advisory Committee. At one point, Taggart asked whether "the Advisory Committee [could] be given guidance as to where to draw the line between projects to be recommended or not to be recommended." A.W. Mackenzie, the Nova Scotia minister of agriculture, pursued the same line of questioning: "What basis does the Advisory Committee use in making such decisions?" For his part, Parker responded rather vaguely that a variety of concerns were taken into account: "The cost per acre, the need for protecting marsh from severe tidal damage, and thirdly the need to protect other facilities in the area."[79]

This was probably not the response that the officials were seeking, as it failed to simplify the process, lacking the precision of a single number. Within weeks of the Moncton meeting, the Advisory Committee agreed that "if each project could be assigned a rating number from a simple empirical formula, this number could indicate the relative value of the project."[80] This led in short order to a meeting that lasted over two days so that a variety of mathematical formulas could be considered. In the end, the committee resuscitated the Haase formula, revised to incorporate an "agricultural rating" for each marsh. This rating, which ranged from zero to ten with zero being the best score, was assigned by the provincial agriculture departments, which provided no explanation of how the scores were determined. With all of the pieces in place, the committee believed that it had found a way "to give in numbers a picture of the marsh as it would be seen by a farmer who lived beside it."[81]

Like Haase's original cost-benefit ratio, the point was to provide a tool, however imperfect, to distinguish one marsh body from another, to permit "the Advisory Committee to arrive at decisions and make recommendations with less difficulty than was encountered in the past [when] this information was not available in concise form." Based on these calculations, the MMRA generated the 1950s version of a spreadsheet listing the marsh bodies in descending order in terms of the likely benefit for each dollar invested. The Wellington Marsh Body (near Grand-Pré) topped the list with a projected $21 benefit for each dollar invested, while the John Lusby Marsh Body (near Amherst), which was largely out to sea, brought up the rear, somehow recording a negative benefit.[82]

These scores, along with the survey maps and other pertinent information, found their way into the files that were adjudicated by the Advisory Committee, whose work — at least in terms of reconstruction projects — was effectively done by the time Parker appeared before a Senate committee in 1961. At that point, 123 projects had been approved, while 33 (roughly 20 percent of the total presented to the committee) had been rejected in whole or in part.[83] Without Parker's evidence, it would be impossible to know how many projects had been rejected because the committee's minutes, which provide a wealth of information about those that were retained, were generally silent when it came to those cast aside.[84]

It is possible, however, to have a peek into the application of the MMRA's tool for rating marsh bodies, with regard to one case that exceptionally surfaced in the agency's minutes. Farmers who owned dykeland in the Centre Burlington Marsh Body, not far from Windsor, had a proposal before the Advisory Committee in 1952 to rebuild dykes and aboiteaux so that 200 acres of land, out to tide since 1942, could be brought back into production. Over the next two years, on three separate occasions, the committee refused to go ahead with the project, following "reports of the Economic and Soil surveys which [had not been] encouraging." To make matters worse, the committee had concerns about "the attitude of the owners and their plans for the utilization of the land," matters that led to the assignment of an agricultural rating of seven, near the bottom of the list.[85] When that rating was incorporated into the MMRA's new formula, the resulting cost-benefit ratio of 1.7 was so low that it appeared unlikely that Centre Burlington would ever see reconstruction work.

To avoid further rejection, just after the introduction of the agricultural rating in March 1954, the owners secured an unprecedented meeting with a delegate sent by the committee, which gave them a chance to have their voices heard. C.E. Henry, who headed the Nova Scotia Marshland Reclamation Commission, met at Centre Burlington with "all owners or their representatives, together with other residents of the community who were interested in the project." Henry was satisfied that this show of solidarity indicated their "intention ... to meet their obligations with respect to a reclamation program." He also learned that the marsh owners were "aware of the survey carried out by the Economics Division and were extremely perturbed by the method used in presenting the report of the survey."[86]

The farmers complained, as had others, that the ratio reflected "what land use would result to the marsh [following reconstruction], based on present conditions in the area, rather than what could be expected to result to the community, if the marsh land were protected and made available to agricultural production." To make this point more concretely, F.T. Beattie, one of the Centre Burlington farmers who sat on the marsh body's executive committee, made a written submission to the Advisory Committee that described how agricultural production had been retarded by insufficient dykeland. There was, for instance, the case of "the Ernest Sanford farm with twenty acres on this tract [which] carries some twenty-five head of cattle and ran out of hay in February of this year. This farm would be able to carry more cattle and benefit greatly with this tract reclaimed." Beattie also cited the cases of other farmers, all of whom guaranteed "payment of any money payable by the Body including ditching costs."[87]

The Advisory Committee considered the Centre Burlington file once more in July 1954, taking into account the new evidence, including a new soils survey that "provided a much more favourable picture." In addition, there was testimony from several witnesses. Henry, based on his meeting with the farmers, reported that "the owners are anxious to have the work done, and give assurances that the land will be used"; and then there was further evidence from the local MP, G.T. Purdy, the only elected official to have had the opportunity to lobby for his constituents before the committee. Purdy made the case that reconstruction would enable farmers to have a balance "between use of woodlot, upland, and marsh." The MP and the other visitors then left so that the Advisory Committee could discuss the project one last time. The crucial moment in the closed hearings came when several members of the MMRA professional staff reported that "based upon the new information that had been received, the agricultural rating of the project could be fairly raised from seven to five." Using this new rating, "Mr. McIntyre [the agency's assistant chief engineer] recalculated the cost-benefit ratio," increasing it to 3.2, which led to approval of an investment of $52,000, mostly for construction of one large aboiteau.[88]

The Centre Burlington case provides a glimpse into the MMRA's procedures, which were created using scientific methods to arrive at a simplified profile of each marsh body. When responding to unhappy farmers who wanted to know why their applications had been turned down, Parker noted that such decisions were taken only after "the Advisory Committee very carefully discussed in as much detail as possible, all the information that was available. This information is as a rule assembled and presented with a great deal of thought and effort. If a project is not recommended for reconstruction, it is because the Committee considers that the expenditure required would not be justified in view of the expected benefits to be derived. It is appreciated that, in some instances, this may be a blow to some individual owners concerned."[89] In practice, however, the Centre Burlington file suggests that the process could be much more complicated, as that particular case included the intervention of the farmers and their supporters, who brought to the attention of the Advisory Committee information that would otherwise have been excluded.

Earlier in the effort to find a single score for each marsh body, Gordon Haase had justified ignoring the views of farmers on the grounds that the agency was trying to create "facts" as opposed to what he called the farmers' opinions.[90] In a sense, he and the experts leading the MMRA more broadly were trying to create a binary that separated their knowledge from that of the farmers. However, just as Parker and his colleagues needed to lean on the traditional knowledge of the farmers to reconstruct the dykes and aboiteaux, they also needed, at least in the Centre Burlington case, to listen to the farmers in determining which projects to support.

On the Ground

Once a project had been approved by the Advisory Committee, it was sent to Ottawa for review by the minister of agriculture. With this formality out of the way, work could begin in earnest, with the Chief Aboiteaux Superintendent (R.H. Palmeter at the outset) in full control, the MMRA's instructions indicating that "authority must be obtained from him before any changes are made in construction methods or in the procedure stated for ordering supplies or any other matter that may come up where a change of procedure was necessary." This centralization of control at Amherst extended to close supervision of such matters as the purchase of material (to be done through the Office of the Engineer), and the labour costs incurred on the site (to be watched over by a timekeeper).[91]

There was nothing surprising or unusual about a newly formed government agency creating a structure that would keep its costs (and public funds) under control, and this extended to the work done on the ground as the MMRA acquired heavy machinery that could be dispatched to jobs as needed, most notably draglines, bulldozers, and ditching machines, doing the work itself instead of having to farm it all out to private contractors.[92] In addition, the agency developed standardized construction practices and carried out the work, particularly for construction of aboiteaux, in its shop at Amherst. As we have already seen, the MMRA was preoccupied with cutting costs through the elimination of aboiteaux, but it also achieved savings by honing its production practices. In 1950, shortly after starting up, Parker explained: "Our shop is fabricating aboiteau sluices and as our production methods are rounded out I expect the cost per sluice will be less than two-thirds of what it was last year." Writing near the end of the 1950s, Parker proudly explained how the MMRA had managed "to develop and standardize designs for aboiteaux, dykes, special dyke facing, and for that matter all structures erected."[93]

To get a sense of how the construction process ideally played out, let's return to the case of Grand-Pré, which was described in work orders written up in late 1949 as "one of the best [marsh bodies] in the province." The MMRA agreed to provide $85,000 to construct 5,000 feet of new dyke and install 14,000 feet of reinforced plank facing. With regard to the latter, the agency was prepared to spend more than it would have in other cases because the Grand Pre Marsh Body had "developed one of the best forms of plank facing, and believe ... that it is more economical than other, cheaper forms of dyke protection." In other words, the agency's practice was shaped by listening to the farmers. The budget also allowed for construction of five new aboiteaux, on all of which the traditional "timber sluice gates [were] replaced by good bronze gates"; the new gates would last longer, making them cost-effective. Finally, the MMRA report included an analysis by its engineers of the work required to reconstruct nearly 1 million feet of drainage ditches, the $43,000 price tag to be split between the province and the marsh body.[94]

Once the broad outlines of the project had been approved, the MMRA and the province's agricultural engineer worked with the marsh owners to iron out logistical details. In February 1950, the executive committee of the Grand Pre Marsh Body received preliminary plans, which were approved with some small tweaks.[95] Parker followed up, informing the marsh body that final plans were still to come, but assuring it that

> it is not our intention to do any work which has not the approval of the Executive Committee. It is our sincerest wish to work with them, co-operate with them, use their judgment, and assist in any way possible. Our organization is new, experienced help is unavailable, and with this in mind it will be necessary for the Executive Committee to be somewhat patient with us, but you may rest assured that we will do the best we can.[96]

To be sure, Grand Pre was a special case, always held up as the model marsh body, where draglines had first been used in dyke construction and where the MMRA's design for aboiteau construction had been developed; and there was also the fact that the Chief Aboiteaux Superintendent, the initiator of that design, had until recently been a member of the marsh body's executive. Grand Pre's special status is further reinforced by the fact that it remains one of the few places where the marsh body still meets regularly. Indeed, I attended its 2019 meeting, which I discuss in the Epilogue. In most places, after the MMRA was created, the local organizations held shorter meetings, met less frequently, or stopped meeting altogether.

The MMRA's centralized construction process borrowed heavily from the practices of the marsh owners, but left them with only a marginal role to play in the new system. Now that the dykes and aboiteaux were the concern of the professionals at Amherst, such matters were removed from the agendas of the marsh bodies, which were left mostly to consider questions related to drainage; and even responsibility for the ditches was now shared with the provinces. In some parts of the region, drainage matters were entirely removed from the purview of the marsh bodies, when the MMRA constructed large aboiteaux and tidal dams across a number of the region's major streams. In a further effort to reduce costs, the agency made the dykes and aboiteaux upstream redundant when tides could no longer enter. In the process, farmers were freed from concerning themselves with routing water toward the aboiteaux.[97]

The decline of the marsh bodies was evident, with the exception of Grand Pre, in the sets of minute books that I was able to examine.[98] It is obviously impossible to claim that these bodies, all from Nova Scotia, were representative of the more than 100 in the province. Nevertheless, I found these particular minute books only because the organizations still exist in some form, allowing the documents to be in the hands of an identifiable secretary, who could let me look at them. Most likely, the marsh bodies that continue to function had greater proprietor involvement in the early years of the MMRA than those that have completely disappeared from view, and yet even these marsh bodies saw their meetings become less frequent and less consequential than when they were still in charge of the dykes and aboiteaux.

In the case of the Falmouth Great Dyke Marsh Body near Windsor, going back to the early twentieth century, minutes typically included reference to dyke repairs, as in 1947, just before creation of the MMRA, when the annual meeting considered both borrowing "$500 for dyke purposes" and acquiring material for aboiteau repairs. However, by 1953, when the MMRA had just finished nearly $100,000 worth of work on the marsh body, the only substantive matter discussed at the annual meeting was a resolution expressing thanks to the agency for its support.[99] In the years that followed, annual meetings were dominated by matters dealing with drainage or concerns to be raised with the MMRA, before the proprietors stopped meeting altogether between 1966 and 1969.[100] Similarly, the

Dentiballis, Victoria Diamond Jubilee, and Bishop Beckwith Marsh Bodies, whose operations stretch back into the nineteenth century, suspended annual meetings for periods of time during the 1960s.[101] Much as the MMRA's own Advisory Committee ceased meeting between 1960 and 1964 because its work had largely been completed, so too did the marsh bodies become less pertinent to the landscape.

The situation in New Brunswick was different, but still spoke to the decline of the marsh bodies. I was not able to unearth a single minute book, to a significant degree because the marsh bodies were effectively put out of business in 1966 by changes connected with the Robichaud government's Programme of Equal Opportunity.[102] Prior to the 1960s, there were significant discrepancies between the services provided by local governments in the various areas of New Brunswick. As a result, wealthier areas were able to provide greater support for education and social assistance because of their superior tax base. In order to give all New Brunswickers a certain standard level of services, all powers over taxation and property assessment were centralized in Fredericton, which in turn provided funding at the local level.[103]

In its 1963 report, the Royal Commission on Finance and Municipal Taxation (often referred to as the Byrne Commission after its chair, Edward Byrne), which laid the groundwork for the Programme of Equal Opportunity, failed to consider whether the new regime would prevent the marsh bodies from continuing their practice of assessing proprietors. Nevertheless, when the legislation affecting the centralization of taxing powers was introduced in 1966, a separate bill was passed to put the marsh bodies in the same category as other local governments. This legislation removed their ability to assess property and levy an assessment on marsh owners, powers that had been a central part of the New Brunswick marshland legislation going back to the late eighteenth century and that were reinforced in the legislation introduced at the time the MMRA was established.[104]

It is impossible to know whether the Robichaud government had anticipated that the changes regarding local taxation would gut the marsh bodies, but that possibility was brought out in the debates over the bill. Claude Taylor, the MLA from Albert County (one of the two counties with dykeland), asked the minister of agriculture how, with the power to tax removed, "the marshland owner [would] be assessed for his share of the work done and how would it be collected?" Taylor then continued:

> Under the old system, I believe there was an assessment and a collection similar to a tax and if a man didn't pay his share on his marshland, then the

marshland could be seized as any other property ... No one should be able to get away scot free, in my opinion, and put a burden on the others. It was never like that before.[105]

For his part, the minister, J. Adrien Lévesque, responded that if money were needed for repair work, then the proprietors could secure the funds through a "gentleman's agreement," but the coercive power of the marsh body, which had previously been able to seize the property of recalcitrant marsh owners, was now gone, a result that did not seem to faze the minister.[106]

With the new legislation in effect, the consequences were spelled out by J.A. Roberts, the director of agricultural engineering for New Brunswick and one of the founding members of the MMRA Advisory Committee. Roberts reminded his deputy minister that the laws put into place in 1949 had been designed to "reinforce the powers of the Marsh Bodies in so far as collection of fair charges for marshland development was concerned." He described a hypothetical situation in which ten owners benefited from improvements to the main outlet drain, but two of them refused "to pay which means that their share must be borne by the other eight owners. In the past, it has been possible for the Marsh Body to legally collect from the two owners who do not wish to participate and their failure to pay resulted in the sale of their land." With this power removed, the two owners who refused to participate would secure a benefit for free, a situation that Roberts found unacceptable, leading him to call for the restoration of the powers to collect assessments either to the marsh bodies or to some other agency.[107]

Roberts's appeal fell on deaf ears, with the result that marsh bodies in New Brunswick ceased to function as institutions responsible for determining what work should be done and how it should be financed. Local self-government of the marshlands effectively ended in New Brunswick, but it seems likely that the bodies there were already in decline before 1966, subject to the same forces that had sapped the vitality of those in Nova Scotia whose minute books I was able to find. In both provinces, little remained for the marsh bodies to do once the MMRA had assumed its most important responsibilities.

Indeed, the MMRA was aware of its role in undermining the marsh bodies. In 1952, only three years after the agency began operations, an internal report bemoaned the inability of the bodies to do the smallest amount of maintenance work, having become entirely dependent on Amherst:

It would appear at the present that the marshland program has done much to reduce marsh owner initiative in respect to marsh protection, and it is a common occurrence to have a marsh owner complain about a leaky aboiteau when all that is necessary is a visit to the sluice at low tide to remove the debris which has been caught in the sluice gate.

Matters had deteriorated to such a degree that the MMRA was considering the replacement of aboiteaux that still had a number of years of service rather than leaving them to the care of a marsh body in their final years, when increased maintenance would be required.[108]

Speaking specifically of the decline in the activities of marsh bodies that had previously been active, the MMRA could have been talking, in the same 1952 report, about the Nova Scotia bodies whose minute books I found: "This loss of initiative is particularly evident in those areas where owner interest was at a high level previous to the program, and it would appear that the task of re-educating these people to maintain dykes, aboiteaux, and breakwaters would be no easy undertaking."[109] In the years that followed, however, there was no evidence that the MMRA tried to educate the marsh owners about any matter, even when it recognized that its own removal of responsibility from the marsh owners had been part of the problem. Instead, there was a constant stream of complaints about how the proprietors could not be trusted, largely a rehash of the commentary that had begun in the 1940s.

A case in point arose in 1953, when the agency distributed its thirty-fifth circular letter to marsh body secretaries regarding the management of their dykelands. It spoke to the MMRA's need to micromanage that it had issued thirty-five of these letters in four years, the most recent of which dealt with erecting fences to protect dykes from damage caused by cattle grazing. The MMRA would provide the material for the fence along with instructions on precisely where it should be installed. After accounting for every possible eventuality, the letter closed: "The foregoing is not meant to be of dictatorial nature but care and cooperation is extremely lacking [from the marsh bodies] in some areas. If the dykes are to last a long period of years they must be cared for and looked after."[110] It seems unlikely that marsh bodies needed to be told that dykes had to be looked after. Nevertheless the MMRA's inclination to hector them was reinforced in an explanatory letter sent by Parker to the agency's district engineers just before the circular was sent out. The director advised that it was best to find "a co-operative member of the Marsh Body's Executive Committee." Expecting some resistance on the ground, he went on: "You and your men will have to be diplomatic on many occasions."[111]

Moving On

Concern over marsh body negligence grew in the late 1950s when Ottawa began talking to the provinces about the return of the dykes and aboiteaux now that the MMRA's work (the large dams aside) was winding down. The initial legislation creating the agency indicated that "the province will assume the operation and maintenance of the work at such time as the [federal] Minister may designate."[112] In turn, the provinces, in their agreements with the newly incorporated marsh bodies, stipulated that they would be passing responsibility for this maintenance work on to the local marsh owners.[113] In that context, a high-level meeting that included the three pertinent ministers of agriculture was held in 1959 at the MMRA offices in Amherst.

J.S. Parker opened the meeting with an assessment of where matters stood, explaining that "by the 1960 construction season, work will have been commenced on all projects which have been recommended by the Advisory Committee [which] has considered practically all projects which have been presented by the Provinces."[114] In fact, however, total MMRA expenditures at the time of that meeting were roughly $15 million, less than half of the final cost of the program when the MMRA closed its doors in 1970. The agency was far from the end of its existence, and yet the talk at Amherst made it sound as if that were the case. Up to this point, the primary mission of the MMRA had been connected to the reconstruction of the dykes and aboiteaux, and that work was now nearing completion, allowing discussion of what came next.

Parker went on to discuss how the provinces needed to design a "well planned maintenance program [because] if maintenance were turned over to the marsh owners, many areas would deteriorate to a state similar to that encountered in 1941 when major rebuilding was necessary."[115] Speaking in such terms, Parker made it sound as if the farmers were *unwilling* to do what was required of them, when in fact many were *unable* to do so as they were experiencing considerable difficulty in financing needed improvement to their lands, most notably in terms of drainage facilities. As the New Brunswick representatives to the 1959 meeting explained: "Ten years ago it was assumed to be reasonable that the marshland reclamation agreements could be followed to the letter, and marsh owners could be assessed for maintenance, whereas today it is believed that this would cripple the economy of those farmers trying to improve their farms."[116]

This question of improvement loomed large as the federal government looked for ways to transfer responsibility for the protective structures. A 1964 MMRA study of "marshland utilization" found that only slightly more than half of all protected dykeland that was suitable for agriculture had been improved, meaning that it had been (or was in the process of being) drained. Based on "intensive field inspections of every lot on every marsh," the study indicated that the problem was greater in New Brunswick, where only 40 percent of agricultural marsh had been improved, as opposed to Nova Scotia, where the figure reached 60 percent, almost entirely because roughly 90 percent of the marshland had been improved in marsh bodies across a stretch extending from Windsor to Kentville, long an area where farmers had relatively high incomes.[117]

To underscore the extent of the problem, the MMRA study showed that through most of the region the majority of land devoted to hay production was unimproved marshland, land that lacked proper drainage but that was being harvested just the same. For instance, the Nova Scotia marsh bodies near the New Brunswick border, where proper drainage was a constant concern, reported that hay was harvested in 1964 on nearly 3,000 acres of unimproved marsh, and on only about half that acreage of improved marsh.[118] This finding was counterintuitive in the sense that discussion about use of dykeland rarely allowed for the possibility that undrained marsh could be productive. Nevertheless, farmers were apparently working those unimproved fields, however inefficiently, as they would have been dealing with standing water. This finding underscored the difficult situation faced by farmers who lacked the resources to drain their lands, leaving them to put their unimproved land to use.

The difficulties of such farmers were on the mind of New Brunswick minister of agriculture J. Adrien Lévesque. In 1965, at almost the same time that the MMRA study on marshland utilization was being released, he observed:

> Farmers in many of our tidal marshland areas appear to be facing a problem that threatens their continued existence ... With the exception of a few small areas [the Windsor-Kentville corridor] where farmers are selling fluid milk, funds were never available to undertake major renovations necessary to bring these lands into the condition where production was efficient and economical.[119]

Lévesque recognized that "the low gross farm income and even lower net income could not possibly meet the demands imposed on it for marshland improvements and at the same time provide for increased mechanization, higher farm expenses and higher living costs." He observed that "where farmers have been able to do the complete rehabilitation job, costs have approached $50 per acre for land forming, drainage, liming, and fertilization." Lacking such funds, most farmers were "unable to improve their lands so that they can expand their operations." He imagined that "many marshland based agricultural communities will undoubtedly go through a painful process of decay over the next ten to fifteen years," in spite of "the public investment [in New Brunswick] approaching $6 million."[120]

With dykeland farmers operating with insufficient incomes, there was, as Lévesque anticipated, a general decline in the number of farms (and the population living on them) and a commensurate increase in the size of the average farm, as departing farmers sold out to their neighbours. These trends were highlighted in a study by G.C. Retson, a federal agricultural economist based in Truro, who focused on the Sackville area of New Brunswick in 1964–65. Retson collected data from twenty-six farms, all of which contained dykeland. He was struck by the low incomes of these farm families, which on average had to live on $1,912 of net income, half of which came from working away from the farm. As Retson put it: "When income is not sufficient to maintain farm capital and provide an adequate standard of living for the family, a movement out of agriculture takes place." Tracking his survey farms back to the mid-1950s, he found that the size of the average farm had grown by roughly 20 percent over the course of a decade.[121] There was nothing unique about this story of agricultural change in postwar Canada. As Helen Parson has noted: "In all provinces east of Manitoba, large amounts of farmland went out of production in the 1950s and 1960s with the most extreme losses occurring in New Brunswick and Nova Scotia."[122] Nevertheless, the situation in the dykeland regions of these provinces received particular attention, given the funds that had been invested.

It was in this context of the fragility of dykeland farming that the federal government was trying to get the provinces to accept responsibility for maintenance. The process started with the meeting in 1959, where both Halifax and Fredericton expressed considerable reluctance, fearing costs they might have to absorb without the participation of the hard-pressed marsh owners. Recognizing this reluctance, the federal minister of agriculture, Douglas Harkness, proposed a gradual transfer that would see the provinces taking charge of maintenance for everything but the large dam projects – recognized as requiring special engineering expertise – by 1962.[123] But 1962 came and went without anything really changing. As J.A. Roberts, New Brunswick's head agricultural engineer, put it: "Repeated attempts to persuade the province to assume the maintenance were successfully delayed by the Ministers of Agriculture until 1965 when it became apparent that the Province could no longer avoid its responsibility under the 1949 agreement."[124]

Ultimately, in 1966, Ottawa and the provinces signed accords calling for the transfer of all of the structures built by the MMRA (including the dams) by 1970.[125] This time the turnover did take place, with the provinces taking full responsibility for maintenance of the protective structures, recognizing that the farmers, who would continue to look after drainage on their lands, could not afford to contribute to the upkeep of the dykes and aboiteaux. The provinces effectively replaced the MMRA, which passed out of existence.

In fact, however, the end of the MMRA, at least with regard to reconstructing the works that the agency had found in disrepair in 1948, had actually arrived in the early 1960s, symbolized in a sense by the departure of two of its earliest employees. In June 1961, Parker left Amherst to take up a senior position with the Department of Agriculture in Ottawa.[126] Only months before leaving, he had an opportunity to review the work that he had supervised in an appearance before the Senate Special Committee on Land Use. At one point in his testimony, he was asked by Senator William Golding whether the $18 million spent by the MMRA to date (roughly half of what the agency would spend before it was done) had been "a paying investment." Parker, who was always highly attentive to detail in the voluminous correspondence that he left behind, was evasive on this occasion, responding: "Senator Golding, down deep I believe that it has [been a good investment]." Not surprisingly, Golding did not leave it at that, remarking: "You think it has?" Trying to be somewhat more expansive, Parker replied:

> I do not know how one can say that an acre of land is worth only so many dollars. I do not know what a group, whether it be private or Government, would say is the limit that they want to spend in protecting or saving land of this type. There must be a limit, but what it is I do not know.

Golding asked one more time whether it was possible "to arrive at some figure of profit or benefits that have accrued from that investment." Parker said it was possible but offered no further commentary, and the exchange ended.[127]

Not surprisingly, Parker's very public exit interview was noticed by the media. *Canadian Business* provided the calculations that Parker was not prepared to provide for the Senators. Dividing the $18.6 million investment by the roughly 80,000 acres that had been protected, "that works out at about $245 an acre, something more than the market price of good farmlands in many parts of the country."[128] However, this calculation misrepresented how farms that included dykeland had functioned for nearly four centuries, combining marsh with upland, where farm buildings were located and other farm operations took place. Speaking to this issue, Parker, in his testimony before the Senate committee, explained how the protected marshland formed "an integral part of an estimated 450,000 acres of farmland."[129] When this information is taken into account, the simple calculation of the cost per acre of the MMRA reconstruction work that Parker avoided and *Canadian Business* provided is significantly reduced.

For whatever reason, Parker was reluctant to make the case for the value of the MMRA investment. He avoided using a single calculation to represent what his agency had accomplished, which was ironic given his penchant for employing such calculations as cost-benefit ratios, engaging in exercises in simplification during his twelve years as director. But the question about the worth of the agency's activities did not go away, resurfacing in a 1967 study by economists Helen Buckley and Eva Tihanyi for the Economic Council of Canada. Their study focused on both the MMRA and the Prairie Farm Rehabilitation Administration, both of which were federal agencies created to deal with water-related issues in rural contexts. The authors assessed the contribution of these programs to the growth of rural incomes, and came to conclusions that were generally positive for the PFRA, even if they found that "the original expectations were overoptimistic." Nevertheless, that agency had made "solid contributions," particularly for farmers who developed large operations during the postwar era.[130]

As for the MMRA, however, Buckley and Tihanyi had little kind to say. Building on data from 1964, they were effectively examining the program as Parker had left it, since the Advisory Committee approved no new projects between 1960 and 1964. While they recognized that the MMRA had protected dykeland that might otherwise have gone "out to tide," the authors wondered "whether the prospects of marshland agriculture could really justify the costs of saving it." While they admitted that they needed more information to give a definitive answer, "the information which is available warrants a tentative answer in the negative." Largely leaning on the studies of the period that were discussed earlier in this chapter, they concluded, as had others, that the benefits of the MMRA reconstruction projects had been compromised by the farmers' inability to carry out "intensive utilization of the marshes that would have required liming, and installation and upkeep of drainage."[131] Given the evidence at hand, Buckley and Tihanyi provocatively suggested that "the abandonment of at least part of the marshland agriculture might have been an alternative worth considering." They conceded that the dam projects might have been worth the expense given the non-agricultural benefits through construction of new highways over some of the region's major rivers, and protection of existing infrastructure, but "in places where agricultural benefits alone had to match the costs, it might have been wiser for the sea to take its toll."[132]

Instead of spending roughly $255 per acre to protect this land (roughly the same amount that *Canadian Business* had calculated after Parker's Senate appearance), Buckley and Tihanyi suggested that "the federal government could have offered a generous purchase price for the endangered lands and allowed their use to the present owners while the old dykes held out." Gordon Haase, the agricultural economist we met earlier, had calculated the value of land in some areas as less than $50 per acre, so Buckley and Tihanyi speculated that the farmers could have been bought out for prices "well above the market price, for a fraction of the $255 per acre protection cost. To the average owner, a cash offer might have meant a substantial incentive to expand upland, or perhaps to finance a move to areas with better employment opportunities."[133]

This reference to the possibility of farmers substituting upland for dykeland suggested that Buckley and Tihanyi were not particularly well informed about farming in the region where marsh and upland complemented each other. Indeed, the authors were roundly criticized for not having talked "to a single person who was at all knowledgeable of the program." New Brunswick's J.A. Roberts was particularly critical, "challeng[ing] these miserable women for the gross inaccuracies and misrepresentations in their report."[134]

More substantively, Roberts refused to accept the premise of the Buckley-Tihanyi report, which was based on calculating (in a further example of simplification) the relationship between the value of dykeland and the cost of MMRA investments. Roberts was an agricultural engineer by training, having served as head of New Brunswick's agricultural engineering service, as secretary of the province's Marshland Reclamation Commission, and as president of the Canadian Society of Agricultural Engineers.¹³⁵ He was an expert working in the employ of the state, but in this and other cases he approached the issue of marshland management as someone close to the ground, displaying mētis as he adapted his professional response in line with local circumstances.¹³⁶ He argued that it was important in the midst of the abandonment of farming by New Brunswickers "to maintain the agricultural economy of an area," even if this meant propping up poor farmers "who could not be expected to adopt more costly farming methods." He insisted that it was the role of government to help maintain "economic farm units. These farmers represent a sizable proportion of the rural population, and the production of these farms provides employment for thousands of rural people." Roberts wanted to "maintain as many people on the land as possible," but without programs such as the MMRA, "we are going to face a problem of each generation having to deal with a new group of rural poor."¹³⁷

For its part, the MMRA meekly responded to Buckley and Tihanyi, noting that "there are areas in which dynamic and positive effects can be seen. Unfortunately, precise documentation does not exist. To acquire it would involve considerable time for field investigation and analysis of data."¹³⁸ Of course, if the data did not exist, then how could they be so certain about the conclusions that might be drawn from it? Much like Parker, the MMRA seemed to feel no need to defend its record, even when its files were full of cases of communities whose land would still be out to sea without its intervention. Roberts was prepared to argue the case, expressing his support for New Brunswick farmers, but perhaps the MMRA officials had already moved on by the late 1960s, their work effectively done. To be sure, substantial work remained to be carried out with regard to improving the now-protected land, but the farmers lacked the funds to do this work, and their organizations – the marsh bodies – which might have lobbied for further provincial government support, had generally become moribund, in part due to the MMRA's own practices.

Only a few months after Parker's departure, Ernie Partridge also left the employ of the MMRA, but under very different circumstances. Ernie had been made aboiteau superintendent for Westmorland County, effectively the marshland area of southeastern New Brunswick east of the Petitcodiac River, because he had a very well defined set of skills that linked the modern practices of the MMRA with older techniques that went back centuries. However, by the start of the 1960s and completion of the reconstruction of the older structures, those skills were no longer required and there was not enough work to justify both Ernie's position and that of the individual who had the same job in Albert County, to the west of the Petitcodiac. As Ernie explained to me: "Well, they started downsizing because they weren't doing near the new construction ... And I had enough seniority that I could've bumped the aboiteau superintendent in Albert County, but him and my boss were bosom buddies, so I said, That isn't gonna happen, so I applied at the [nearby Dorchester] penitentiary." Ernie unhappily moved from one federal institution to another, leaving a job that he loved for one that he described as "the most unthankful position on the face of this earth. At the end of eight hours, you'd come home and you knew you never accomplished one thing." In the end, however, Ernie had little choice because, as he put it, "I had seven little fellas in this house, runnin' round, that needed to be fed."[139]

Not that it would have made him feel any better, but Ernie Partridge was not alone in leaving the employ of the MMRA in the early 1960s. A 1964 report on the future of the agency noted that "with the work for which it was originally established almost completed, the organization is somewhat smaller than previously, now with a staff of approximately thirty-nine."[140] This report was commissioned by Ottawa as part of a plan that would lead to even further reductions to the staff at Amherst, as the MMRA became integrated into the Agricultural Rehabilitation and Development Administration (ARDA), a federal agency created in 1961 to work with the provinces to "improve the use and productivity of resources in rural areas."[141] There was a certain logical connection between the two agencies, both of which were focused on "rehabilitation," leading in 1963 to "the MMRA [being] designated as the operating arm of ARDA in the Atlantic Provinces," providing expertise on soil and water issues that extended far beyond the dykelands.[142]

In the years that followed, while engineering services remained in Amherst, the MMRA administrative jobs were shifted to Moncton, leaving local officials to lobby against a move that would have consequences for the region. A flurry of telegrams were sent to the minister in Ottawa, emphasizing how the loss of employment would hurt an already fragile economy, and a formal submission from town leaders stressed how "the MMRA staff members have been residents of Amherst since 1949 and are active in all phases of community life. The majority of these employees are homeowners ... The larger percentage of these employees have expressed desire to remain as residents of this community."[143] Phrased in this manner, it made the case that the MMRA professionals, some of whom hailed from elsewhere, were embedded in the community, providing knowledge that had become, in a sense, local.

By the early 1960s, with the reconstruction of the protective works completed, J.S. Parker had moved to Ottawa, Ernie Partridge had moved on to employment at the penitentiary, and numerous other MMRA employees had been let go or transferred to Moncton. And then there were the marsh owners who did not have the means to make their newly drained dykeland profitable and had left farming altogether. This first phase in the MMRA's history had returned the landscape to something approaching what it had been before the collapse of the hay economy in the early twentieth century. But the results of this work benefited only those with sufficient resources to improve their lands. The melding of traditional and expert knowledge had achieved significant results, but for many farmers, the reconstruction of the dykes constituted a hollow victory, as they found themselves looking out at unimproved fields.

PART 2
Third Nature

3
Dam Projects

IN THE FALL OF 1952, marsh owners met in Sackville, New Brunswick, to discuss a project that had not even been officially announced. The meeting was called by Laurie Anderson, a local farmer with 130 acres of dykeland. He opened by explaining why he had called this group together: "Man who should be in the know told me that plans are ready and that the Tantramar River will be dammed next year." Anderson explained that he had "made a commotion at a meeting [of the marsh body] at Point de Bute recently to call a meeting to discuss the damming – no one would second the motion so I called this meeting."

There were numerous assemblies on a wide array of topics during the years that the Maritime Marshland Rehabilitation Administration was in operation, but this was the only one called spontaneously by an individual rather than an organization such as the MMRA or one of the marsh bodies. Indeed, we know that it took place only because someone, most likely one of the MMRA's officials in attendance, took careful handwritten notes about "Laurie Anderson's Meeting" that found their way into the agency's surviving files.[1]

At the time of the meeting, the MMRA was still in the early stages of thinking about building large structures near the mouths of the region's major rivers, such as the Tantramar, to block the movement of the tides and to allow the existing dykes and aboiteaux upstream (which would become irrelevant) to deteriorate, resulting in considerable savings. In moving from the restoration of older structures to the construction of tidal dams, the MMRA found itself in the mainstream of postwar development across

North America, which looked to a muscular control of nature, specifically of water. The agency shared the perspective that took hold, for instance, in the American West. Donald Worster has shown how the US government's introduction of a program to aggressively control water in the face of drought "was not an acceptance of and adaptation to [arid conditions], but technological mastery over that ecology."[2]

The idea of damming rivers that flowed into the Bay of Fundy had first been expressed in the early 1940s as the federal government was looking into emergency measures to stop the decline of the protective structures and the resulting washing away of dykeland. E.S. Archibald, the director of the Dominion Experimental Farms, commented frequently on how the pertinent governments needed to consider building "permanent dams and large aboideaux at the mouths of these tidal rivers," in the process "eliminating possibly hundreds of miles of dykes."[3] In this formulation, Archibald used the terms "aboiteau" and "dam" interchangeably, as would many others in the years to follow, and while both types of structures could achieve similar goals, they did so in different ways, with different consequences.

In the pre-deportation era, aboiteaux had been constructed to allow fresh water to drain out at low tide, while preventing the entry of salt water at high tide. In a 1953 report, the MMRA explained how, starting in the seventeenth century,

> when marshes were brought in from the tides, aboiteaux were placed in all creeks which had been created by the action of tidal water. As all ditching was carried out by hand work, it was no doubt easier and cheaper to build the aboiteau than to carry out the ditching necessary to drain an existing creek to another aboiteau site.[4]

Over time, as technology evolved, increasingly larger streams were blocked by ever larger aboiteaux, and by the time the MMRA was on the job, some of the major tidal rivers of the region were targeted for obstruction by such structures.

In this regard, the agency set out in the early 1950s to construct an aboiteau across the Isgonish (today the Chiganois) River just west of Truro. George Frail, one of the agency's engineers, noted:

> As the river winds through the marsh land for over a mile it was obvious to the first Engineer who visited the area that the logical procedure would be to construct an aboiteau in the main river. This would eliminate the neces-

sity of constructing several miles of dykes along the river banks with the many expensive drainage outlets [aboiteaux]. The land could then be used right down to the water's edge and drainage would be greatly simplified.

To underscore this point, Frail included a map that showed the twenty-seven aboiteaux that might be eliminated by building a structure near the point where the Isgonish River flows into Cobequid Bay, the easternmost extension of the Bay of Fundy.[5]

While Frail's aboiteau, completed in 1955, was much larger than those that had been constructed by marsh bodies prior to the establishment of the MMRA, it was still an aboiteau, serving the same function as the hundreds of such structures that had been constructed over the previous 300 years, all of which were designed to expel fresh water from the inland side of dykes.[6] But there was something very different being proposed for the Tantramar River, prompting the alarmed response by Laurie Anderson and his neighbours.

In this particular case, the MMRA was not proposing just any obstacle to the flow of the tides upstream, but rather the construction of a dam. Such structures formed a fixed barrier between fresh and salt water, in the process creating large reservoirs (often referred to as headponds) on the inland side. Unlike aboiteaux, dams prevented water from escaping, except when it was necessary to release it to keep the depth of the reservoir below the height of neighbouring dykeland to allow drainage from the fields. In addition, dams changed the dynamics of tidal rivers as vast quantities of silt that had previously washed up and down the river twice daily, now settled out downstream from the structure, frequently reducing both the width and the depth of the stream. Besides the more visible changes brought about by dams, there were others that were harder to see but important just the same, such as the complications created for the movement of fish. This was particularly the case for anadromous fish such as salmon, which are born and which spawn in fresh water but spend most of their lives in salt water. In short, the dams transformed both the region's landscape and its ecology. Following first nature, in which the tides moved freely across the marshes, and second nature, created by the dykes and aboiteaux, here was third nature.

Before the MMRA wound up its work in 1970, it had built tidal dams across the Shepody, Tantramar, and Petitcodiac Rivers in New Brunswick, and the Annapolis and Avon Rivers in Nova Scotia (Map 2).[7] These dam projects are the focus of this chapter and much of the next, as they

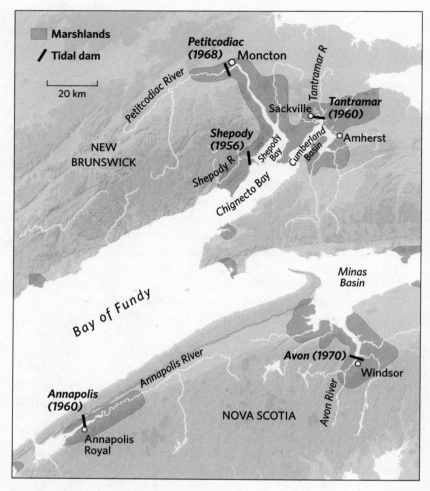

MAP 2 MMRA tidal dams. *Map by Eric Leinberger.*

constituted a significant initiative not only for the region but also for the regulation of water more broadly. Indeed, when the Shepody Dam, the first of these projects, was completed in 1956, it was "the first major tidal dam constructed on the North American seaboard."[8] And as a new initiative, the dams frequently generated opposition, as was the case with the proposal to dam the Tantramar River that led to Laurie Anderson's meeting.

Recognizing that the river would be cut off with the construction of a dam, some farmers in the Tantramar area wanted it to be located as far upstream as possible to allow continuation of the practice of tiding, which had a long tradition in the region. As we saw in Chapter 1, farmers would occasionally open the gates on aboiteaux or create a breach in the dyke to allow silt-laden salt water to settle on their lands, in the process adding nutrients that they believed would make their drained marsh more productive. If the Tantramar Dam were located at the mouth of the river, tiding would be impossible as the tides would be blocked, but moving it further upstream, aside from limiting the number of aboiteaux that would be eliminated, would result in a reservoir that was too small to operate successfully. As Ron McIntyre, the MMRA's assistant chief engineer, put it at Anderson's meeting:

> Physically an aboiteau [meaning dam] could be built there [upstream], however it appears likely that the size of the storage basin in the river above that point would not be sufficient to prevent flooding of the land during freshets [the freshwater run off]. Whereas further down the river, sufficient storage is available ... because runoff during one tide cycle would not be large enough to fill the reservoir.[9]

The reservoir issue touched on tiding, but it was the challenge to the practice itself that prompted the strongest intervention from a marsh owner against the dam project. Ern Estabrooks insisted that "marshland is probably the most fertile land lying outdoors. However, the more often we plow and crop the quicker the fertility becomes exhausted. It has been found that the most reasonable way to replenish them is by tiding. A dam will forever cut off this opportunity." On this occasion, J.D. Conlon, the MMRA's chief engineer, did not feel the need to respond specifically to the issue of tiding, taking the easy way out by noting that "plans have not been finalized nor has a decision been made in respect to the damming of the Tantramar River."[10]

While Conlon held his fire on this occasion, tiding was generally held in low regard by MMRA officials, who viewed it as representative of the farmers' primitive practices that had created the situation making the MMRA necessary in the first place. Indeed, one of the attractions of dams was that they would remove control over dykeland protection from the owners. As MMRA reports would indicate on several occasions (using the identical expression), with a dam

> responsibility for maintenance of the structure would rest with a central authority. In the case of dykes and aboiteaux the responsibility for maintenance would rest with several individual groups. Experience has proven that

when the latter is so, particularly if the groups consist of owners dependent on one another, satisfactory results are not always maintained, much to the detriment of the community concerned.[11]

This statement was a particularly high-modernist view of the utility of dams, touting the superiority of centralized, expert control over dispersed control in the hands of farmers leaning on their local knowledge. As we saw in Chapter 2, these two forms of knowledge did not have to be at odds with one another. Indeed, the application of mētis, the incorporation of local knowledge into expert thinking, was at the heart of the reconstruction of the dykes and aboiteaux. However, a different dynamic was at work in the case of the dams, which were large, expensive engineering projects that could not, from the MMRA's perspective, be aided by the practices of the marsh owners.

Even more problematic, however, was the agency's failure, in terms of some of these projects, to alter its practices when circumstances dictated. As James Scott put it, it was important for high modernists to "know how and when to apply rules of thumb in a concrete situation," but this was not always the case, with significant environmental consequences.[12] Accordingly, we will see that practices that worked for the early dam projects were not appropriate for some of the later ones. Tina Loo describes how such projects worked best when the engineers exhibited a certain "high modernist local knowledge," combining "the impulse to control and dominate nature, on the one hand, and to live humbly and harmoniously with it on the other hand."[13] In the case of the MMRA dams, however, particularly the last two dams constructed across the Petitcodiac and Avon Rivers in the late 1960s, the agency seemed to have lost sight of the second part of this equation.

That said, it is important to avoid viewing these projects one-dimensionally as agents of environmental degradation. As Loo has explained, "we should pay attention to what is created as well as what is degraded or destroyed." By the time the MMRA closed down in 1970, nearly 33,000 acres of marshland had been permanently protected by the agency's five dams, although in a number of cases it might have provided the same protection through other, sometimes less expensive means. In addition, the dams provided protection for infrastructure across the region, including railways, highways, and public buildings. In fact, a number of the dams were viable economically only because government agencies that needed to build roads over the rivers were happy to share the costs with the MMRA, in the process turning several dams into causeways.

And finally, there were those headponds that, as we will see in the following chapter, were demonized by some as a scar on the landscape but that provided opportunities for others, most notably with regard to the structures built across the Petitcodiac and Avon Rivers. In the former case, a suburban community emerged along the banks of what residents called Lake Petitcodiac, naturalizing the headpond in the process; in the latter, the community largely embraced Lake Pesaquid because of the recreational opportunities it provided. As Loo put it, "acknowledging [such creations] does not diminish the destruction that occurred, but it highlights how high modernism benefitted some people and not others, and recognizes how in the process of destroying certain landscapes and lives high modernism also reshaped and created the terrain on which we subsequently must search for solutions."[14]

Dam construction constituted a distinct moment in the history of the MMRA, as even the agency recognized. Beginning in 1961, it produced annual statistics that distinguished between the "acreage protected by standard MMRA protective works" (including aboiteaux of all sizes) and that which had been secured by "major tidal dams." The timing of these statistics reflected the fact that by the early 1960s the agency had finished the reconstruction projects described in Chapter 2. Even though serious discussion of dams had begun in the early 1950s, as reflected in Laurie Anderson's meeting, only the dam at Shepody was completed during that decade. At the start of the 1960s, dams were responsible for only 10 percent of the protected dykeland, a figure that steadily increased, reaching 40 percent a decade later, in the process pushing the total area of protected dykeland to over 80,000 acres.[15]

In the transition from reconstruction work to dam projects, the voices of the marsh owners, which were marginalized as the MMRA rebuilt the dykes and aboiteaux, were largely silenced. Laurie Anderson's meeting was a moment when the concerns of farmers in the face of significant changes to their landscape could be heard. There would be a few other such moments in the early 1950s, but the MMRA quickly tired of consulting the marsh owners so that the record that follows is largely constructed from the evidence provided by an agency committed to creating third nature.

The Test Case

Even before the MMRA was formed, the leading candidate as the test case for dam construction was the Shepody River in southeastern New Brunswick.

In 1943, at the high-level meeting that set the stage first for emergency repairs and subsequently for creation of the MMRA, politicians, bureaucrats, and agricultural experts considered the relative advantages of "dyking all along stream banks" versus placing "aboideaux across river mouths." In the discussion that followed, only one specific example was provided that fit into the second category. New Brunswick's deputy minister of agriculture "advised regarding conditions at Shepody where a 1500 foot [tide] control would replace twenty-five miles of bank dykes."[16] When the idea for blocking tidal rivers resurfaced six years later at the first meeting of the MMRA's Advisory Committee, Shepody was again explicitly mentioned, and only weeks later J.S. Parker, the founding director of the MMRA, began the process of carrying out a feasibility study for this project.[17]

Like many other dykeland regions, the Shepody watershed had seen much better days by the time talk about building a dam at the river's mouth began. A study commissioned by the provincial government in the early 1950s indicated that "the present condition of the area is largely due to a decline in farm prices and income during the thirties which resulted in farmers being unable to maintain the dykes protecting the marshland."[18] The full impact of the deterioration of the protective works along the Shepody, in the Harvey Bank Marsh Body, was driven home by a report commissioned by the Dykeland Rehabilitation Committee, which looked after emergency repairs in the 1940s:

> There are three aboiteaux in the body. These aboiteaux are suffering at the present time as tidal water rushes through the gaps in the dyke and the most of this must be discharged through the aboiteaux which were not constructed to handle this flow of water. The result is considerable washing on the inside face of the large aboiteau [which] will be much weakened and may [soon] be out. Replacing such an aboiteau is beyond the means of the farmers.[19]

This description could also have applied to other dykeland areas at the time, but what made the Shepody situation distinctive, and as a result attractive as the site for the first dam, was the large area of marsh that was out to sea and that could be recovered only through construction of a structure at the river's mouth. There were roughly 5,000 acres of marshland here. Three thousand acres could have been protected from the tides through either reconstruction of the dykes and aboiteaux or construction of a dam. The remaining 2,000 acres, however, were potentially arable marshland that "were out to sea and would remain out to sea or flooded by fresh water if an attempt was made to protect the area [solely] with dykes and aboiteaux."[20] In other areas, such as the watershed of the nearby Tantramar River, the area of land that could be protected *only* by building a dam was relatively small. Individuals could debate the pros and cons of building a dam as opposed to reconstructing the dykes when the same area of marshland was in question, but at Shepody no such debate was possible, as the dam would bring into production large expanses of marsh that would, according to Parker, otherwise be "impossible to reclaim."[21]

In this context, the Advisory Committee met to approve the Shepody project in June 1951, at a two-day meeting attended by senior officials from Fredericton, including the minister of agriculture. The committee members asked few questions about the feasibility of a project unprecedented in this part of North America, in which a fourteen-mile tidal river would be reduced to one less than two miles long, since the proposed site of the dam was relatively close to the mouth of the Shepody. There was not much discussion either as to whether the project was worth the investment of roughly $1.5 million, which at the time was roughly the total of all MMRA projects since its inception.[22]

Instead, the brief discussion of the Shepody project focused largely on the marsh owners, and whether they were likely to improve their farms if 5,000 acres of dykeland became available. Such a discussion was consistent with MMRA concerns in the early 1950s that farmers, leaning on their local knowledge, would fail to put to good use land protected from the tides through the agency's reconstruction of dykes and aboiteaux, or in this case through construction of a dam. As a result, in the months leading up to the Advisory Committee meeting that gave the go-ahead for the dam project, the MMRA and its provincial partners gave considerable attention to the role of the proprietors.

For instance, in March 1951, the agency organized a meeting attended by roughly forty farmers representing "upwards of seven communities," so that it could gauge the engagement of the farmers in looking after their farms. Parker explained at the start that "the investment of the Federal Government is of no use unless land is used." The meeting then considered a variety of practical questions, with one of the owners observing that "if farmers were assured of no tide, they would feel more like using land to better advantage." The transcript of the meeting suggests that the mood in the room was positive, leading the farmers to "agree without a dissenting vote to a proposal to erect a dam across the Shepody River." Along the way, the marsh owners also supported a proposal that, following completion of the dam, each of them should "spend approximately $15 per acre for main drainage," seemingly responding to MMRA concerns that they would not hold up their end.[23]

In spite of such signs of support, in the months that followed, various reports reflected concern that action was needed to guarantee that the owners would do the right thing with their land. Along the way, officials paid little attention to whether farmers could actually afford a course of action that, as we have seen, had proven too costly to improve land protected by dykes and aboiteaux. One provincial study suggested that "the added security guaranteed by the construction of a dam should help to impart a feeling of safety that would encourage farmers to spend money on improvements [to their lands]. This guarantee will not be present if the lands are protected by a system of dykes and aboiteaux as they have been in the past."[24] Nevertheless, the same study went on to outline actions to ensure the farmers' proper attention to their lands, calling for the

> appointment of a special Agricultural Advisor who will be stationed at Albert [near the site of the dam]. The individual to fill this post must be carefully selected and trained. His duties are to work with the farmers of the area. He should be of the better type of "super salesman" who sells ideas, not merchandise, and does it in such a way that the farmers think it was their own idea ... We feel that it is essential to create a desire for improvement before any progress can be made.

The study also proposed that the Agricultural Advisor should be assisted by a "Community Planning Board" made up "of interested and progressive farmers ... to help promote useful developments and better land use."[25]

It is not clear whether either the Agricultural Advisor or the Community Planning Board was ever appointed. Nevertheless, the sentiment expressed with regard to the farmers suggested that the Shepody project had some of the characteristics of a classic high-modernist initiative in which both the landscape and the people who inhabited it needed to be "rehabilitated," to use the expression that was popular in the postwar period in a variety of settings.[26] A further study on "the rehabilitation of the Shepody area" pointed out that "it would be wrong to create the impression that as soon as the dam is completed, every drainage ditch will be cleaned and every acre reconditioned ... The complete rehabilitation of the area may require ten to fifteen years in order to allow for the natural replacement of individuals who cannot or will not use the land."[27] In short, the farmers needed to be fixed at the same time that the tides were being held back, in this case by a dam.

Following these studies, when the Shepody project was considered by the Advisory Committee, James Anderson, a member who belonged to the Anderson family of marsh owners from Sackville, asked about the "sentiments of the members of the Executive Committees of the marshes involved."[28] This was an interesting question since the marsh bodies were supposed to be the vehicle for advancing the collective interests of the farmers, but there was little evidence of the participation of these bodies in the process. When farmers attended the public meeting described earlier, J.K. King, the province's deputy minister of agriculture, "assured those present" that they were speaking "as individuals – not as representatives of the Marsh Bodies."[29] This observation was consistent with the MMRA's actions that, as we saw in the previous chapter, had the effect of sapping the authority of the marsh bodies.

As for the response to Anderson's question, the professionals at the Advisory Committee meeting, from both the MMRA and the provincial department of agriculture, suggested that consent had been obtained from the marsh bodies. While Ron McIntyre, the agency's assistant chief engineer, claimed that "the committees of the Hopewell Hill, Upper Dyke, and Germantown Marsh Bodies were in favour of the dam in principle," other responses were less convincing. J.A. Roberts, New Brunswick's lead agricultural engineer, assumed that the Harvey Bank Marsh Body, close to the dam site, was onside because "some men at Harvey had suggested one large dam when he had first visited marshland in that vicinity some seven or eight years ago," which is to say even before the MMRA had been created. For his part, K.L. Langmaid, a soil engineer based in Fredericton, offhandedly observed that "the men in the Riverside area [also near the dam site] are in favor of the dam," without explaining how he knew this to be true.[30] In any event, Anderson did not press the issue, and the dam project was unanimously approved.

The Shepody project moved toward construction with the signing of a contract in 1953. Over the years that followed, every step in the process was carefully documented in the MMRA photo archive: from tests in the agency's hydraulic lab at Amherst, to the various phases of construction, and finally the completed dam (see Figures 3.1 to 3.4). The project also generated significant interest in the local press, much of it expressed in heroic terms. There was, for instance, the report in 1954 from the *Moncton Daily Times*, which described the work of

> a mobile labor force, working diligently around the clock and through the seasons ... The army of mechanized human beavers has been confronted

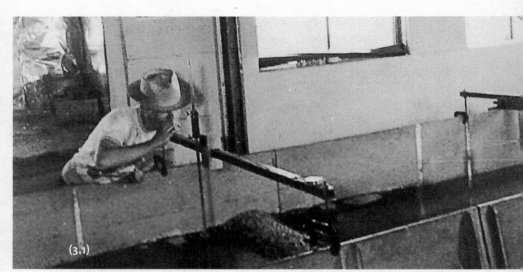

(3.1)

(3.4)

Dam Projects

FIGURES 3.1–3.4 (*clockwise from top left*)
Construction of Shepody Dam: Shepody model test in MMRA lab, July 1953; Completed concrete works, October 1955; Completed dam, showing silting, May 1956; Shepody River, 1962. *New Brunswick Department of Agriculture.*

with substantial hazards for both men and machines ... At peak strength its numbers have reached nearly sixty men and as many pieces of equipment. Outmaneuvering swift running tides to gain vital extra minutes of work, battling ice floes and other obstacles of nature.

Viewing the closure of the river, the defeat of nature, in military terms, the article ended by noting how "this force has set the fall of 1955 as completion target."[31]

As it turned out, the dam was completed a bit behind schedule, the river being closed in March 1956, leading the Advisory Committee to note in its minutes a few weeks later how "the tide could now be shut out of the reservoir, and kept off the marshes."[32] At the very end of the same meeting, New Brunswick's lead agricultural engineer made some remarks about the new landscape that had been created. J.A. Roberts observed that "he would not be surprised if a petition for holding the reservoir at some elevation above normal low water flow would be received. It was conceivable that in the future, agricultural conditions might change so a reservoir of fresh water for irrigation would be desirable."[33] Here was the MMRA, for the first time, contemplating how it could contribute to providing – as opposed to eliminating – water, and it was doing so in the context of a headpond, a feature of the new landscape connected with dam construction.

The completion of the Shepody Dam also quickly led to another significant change in the landscape, with the buildup of silt downstream from the structure, as shown in the last two photos above, the first of which was taken only weeks after closure of the river. This situation is shown more dramatically in the following image from 1962, where only a narrow channel remains open below the dam (to the right in the photo). This situation was caused by the end of a process that for millennia had seen half a million tons of sediment wash up along the full length of such tidal rivers twice daily, and then wash back out. C.A. Banks, the MMRA's chief of technical services, explained the situation in "normal" tidal rivers: "As the tide ebbs, the freshwater run-off, which also carries material in suspension, transports some of this material to the river mouth ... Over a period of years, the tidal/run-off interaction reaches a state of equilibrium."[34]

A dam completely changed the dynamics, creating a downstream flow that was significantly reduced as the tidal section of the Shepody River shrank from fourteen to two miles in length. As a result, silt settled out immediately downstream from the dam, reducing both the width and depth of the river. Banks reported that over the first year following closure of the Shepody River, "at a point 200 ft. downstream of the face of the dam," the silt buildup "amounted to 15.5 ft., or just in excess of .25 inches per tide."[35] In 1958, J.D. Conlon, the MMRA's chief engineer, reported that "the area below the [Shepody] Dam was still silting in rapidly."[36]

At the meeting of the Advisory Committee that discussed the Shepody project, Parker observed that "one point which should be considered before a structure of this type were erected would be the possibility of the existing channel silting up which would pose a drainage problem in years to come. It is our general opinion that this is not apt to pose a problem on this stream." Speaking in such terms, Parker was only considering the buildup of silt above the dam, which might block drainage channels that still needed to lead water from the marshland. He minimized this upstream problem because "the silt from the tide will be kept out." Exhibiting the tunnel vision that often came with high-modernist projects, Parker was silent when it came to the changing dynamics in the river below the dam – but perhaps it did not matter because the MMRA was focused on the land above the structure.[37]

Good Tidings

The Shepody Dam moved to completion without any of the problems that would mark the four subsequent ones. Its cost was easily justified by the large amount of marsh that could be protected only by a dam, a factor that encouraged farmers to accept its construction without reservations; and if there were concerns about the impact of the dam on the landscape, they never surfaced. By contrast, the next dam to be constructed, across the Tantramar River, had already generated controversy even before its approval by the MMRA, as was evident at Laurie Anderson's meeting in the fall of 1952. The issue at the meeting was the question of tiding, which inspired local marsh owners to stand up against the plans of the MMRA in a way that was not seen in connection with any other project in its history.

Tiding was an environmentally sustainable activity that had a long tradition in the Tantramar watershed, stretching back to the early nineteenth century. It was embodied by Toler Thompson, a local farmer who was responsible for constructing a canal that carried tidal silt to areas far from the river that would otherwise have been bog, a type of wetland that (unlike marshland) is nutrient-poor and without any connection to a watercourse. Thompson's initiative differed from the practice that went back to pre-deportation times, in which Acadians (and their English-speaking successors) allowed the deposit of silt on dykeland. Nevertheless, his efforts were so successful that his "name is justly held in high honour in the vicinity," and became an inspiration for local farmers in the region who jealously guarded what they saw as their right to be able to continue to tide, to build up both marshland and bog.[38]

While Tantramar farmers felt a strong, positive connection with the tradition of tiding, professionals who worked for or with the MMRA viewed it as representative of outdated local knowledge. This point of view had frequently been expressed during the 1940s, particularly by E.S. Archibald, the director of the Dominion Experimental Farms and the leading advocate for experts to take control from farmers in the reconstruction of the dykes and aboiteaux.[39] In the early 1950s, just as the MMRA was beginning to promote the Tantramar Dam project, Parker received a telling letter from his close collaborator, J.A. Roberts, who was writing about a proposal for dykes to be broken in the vicinity of Sackville (near the probable site of the dam) in order to encourage tiding. As Roberts put it, "I do not see that the Federal Government has any moral legal obligation to undertake these works."[40]

Roberts's first choice of words was telling, and reflected what James Scott called the high-modernists' interest in advancing a certain "visual codification ... Their vision required a sharp and morally loaded contrast between what looked modern (tidy, rectilinear, uniform, concentrated, simplified, mechanized) and what looked primitive (irregular, dispersed, complicated, unmechanized)."[41] MMRA leaders were put off by the idea of tides washing up on arable land, and were very much attracted to the image of the endless, uniform hayfields of the Tantramar marshes (see Figure 3.5) and the sharp lines that might be established between salt and fresh water through construction of a dam.

Parker took the first steps regarding the Tantramar Dam project in early 1950 when he wrote to his deputy minister explaining why it was necessary. From the start, Parker recognized that there would be opposition from farmers, dismissively referring to how "the local people would be content to have the old dykes maintained by hand, thus providing a small source of income for certain periods of the year." In fact, however, there was little evidence that anyone made money from the old system of maintaining dykes and aboiteaux, which generally sapped the limited funds of farmers and led to creation of the MMRA. Parker went on to describe how opposition to the dam was fuelled by support for tiding, which would end should the dam be constructed at the mouth of the river: "I do not feel

FIGURE 3.5 Hayfield in Tantramar Marsh. *Mount Allison University Archives, Robert J. Cunningham Fonds, 7922/4b/29.*

consideration toward it [tiding] is warranted at this time. If tiding were deemed desirable to reclaim new land thirty or forty years from now it would be possible to make this a separate project and install tide gates for tiding purposes across the neck of land adjacent to the proposed sluice."[42]

At Parker's insistence, federal funds were provided to carry out a feasibility study for the Tantramar project, which was tabled for discussion by the MMRA Advisory Committee in early 1953, but whose existence was sufficiently known in the preceding fall to generate the opposition on display at Laurie Anderson's meeting.[43] The feasibility study echoed many of the arguments used to justify the Shepody Dam, repeating word for word the belief that a single structure was better than leaving responsibility for protection from the tides with the farmers. At the same time, however, the study recognized that roughly the same amount of land could be protected by building a dam as by reconstructing the dykes and aboiteaux, unlike the situation at Shepody, where a dam would greatly increase the supply of arable marshland. This fact overshadowed the debate about tiding because the MMRA could not draw the marsh owners in by arguing that the dam would provide more land to local farmers than the existing structures. The feasibility study, published after Laurie Anderson's meeting, anticipated "protests by individuals or groups against the building of this structure," and its authors advised the New Brunswick government "to obtain from the owners, by means of local plebiscites, their reaction to the proposal to erect a large structure."[44]

The feasibility study, and more specifically its references to tiding and the need to respond to local discontent, provided the focus for a two-day meeting of the Advisory Committee in early 1953 at which a number of soil specialists were on hand to provide their expertise; as for the marsh owners, they were conspicuous by their absence. Since most of the advantages of a dam had already been expressed with regard to Shepody, the discussion quickly turned to *"tiding"* (both the quotation marks and italics were the MMRA's emphasis), described in the meeting's minutes as "the most debatable problem involved in the project." With regard to allowing tides to occasionally run across marshland, the committee found that the practice was not "economically feasible, because of the comparative costs of tiding and of adding soil amendments (such as lime and other chemicals)." As for the practice of tiding to build up bog areas inland (following Toler Thompson), it concluded that "the benefits may be only of short duration due to consequent settling after a few years."[45]

Since the marsh owners did not participate in the MMRA's discussions, the community remained concerned, leading the New Brunswick Marshland Reclamation Commission to authorize "a meeting of the marsh owners of the Tantramar area to discuss and vote on the proposed Tantramar Dam," an action that was subsequently endorsed by the MMRA.[46] The meeting held in Sackville in late April 1953 was attended by "about 125 marsh owners and interested parties." It was chaired by J.A. Roberts, who soon lost control when Laurie Anderson took the floor, insisting that there needed to be a better process than having decisions made behind closed doors by the MMRA. Anderson called for the creation of "an organization of marshes at the head of the Bay of Fundy, with a committee of about three to advise the MMRA on all projects," in essence a grassroots version of the Advisory Committee. He then segued into the substance of his concerns, proclaiming with regard to tiding: "We do not want to shut off a God given heritage for all time," following which Roberts needed to call for order.⁴⁷

When everyone had calmed down, Parker responded directly to Anderson, providing a moment when expert and local knowledge were both represented on the same stage. The director of the MMRA pointedly referred to Anderson's claim that tiding was a "birthright ... It is true that you will not be able to tide. But it is our heritage to use the marsh. So we must plow and lime the soil to use the land economically." Roberts then intervened, questioning what farmers would do if they tided their lands and had to deal with a residue of salt for several years, to which Anderson responded: "Men can cut the salt marsh for feeding while the tiding is going on." This response rejected the premise of MMRA activity that salt marsh could serve no productive function, but as we saw in the previous chapter, there was a significant amount of unimproved marshland that was producing hay in the early 1960s. Effectively, Anderson was touting the value of first nature and challenging the idea that the only solution was to keep the neat division between water and land. As the meeting drew to a close, the farmer dug in his heels: "God almighty gave us a heritage and we have not the right to dam the river ... The day will come when the land will need tiding and there will be no way to get the tide ... Let's organize the marsh owners tonight."⁴⁸

At the end of the meeting, a vote was taken, and the vast majority of the 63 farmers on hand approved the dam. However, Roberts was unsure about what could be concluded from a vote by roughly one-quarter of the 255 marsh owners who would be affected by closing off the river. In order to establish whether "the vote represented a true cross-section of marsh owners," Roberts asked those who participated in the plebiscite to indicate how much land they held. It was found that small landowners had been under-represented in the vote, as they constituted only about one-quarter of those who participated in the plebiscite but two-thirds of all marsh owners. As for the overrepresented large landowners (holding over 100 acres), they voted almost unanimously for the dam, giving substance to Roberts's concerns about what could be gleaned from the process.⁴⁹

The vote had not really changed the situation on the ground, and in June Laurie Anderson called yet another meeting of marsh owners because a "number of them felt that the installation of an aboideau [a dam at the mouth of the Tantramar River] would not be in the interests of many of the farmers in the area." He circulated a petition against construction of the dam, and urged the farmers to create "an association which would be able to intelligently interpret the desires of farmers and present their views to both the provincial and federal government."[50]

Perhaps it was only coincidental, but on the same day as Anderson's most recent meeting, the New Brunswick Marshland Reclamation Commission authorized J.H. King, from the provincial department of agriculture, to carry out a "personal canvass" on the subject of the dam. Essentially, he went from farm to farm seeking out farmers' views, but once again failed to obtain the results that the commission (or the MMRA) was hoping for. On this occasion, each of the 255 farmers was contacted, but while 45 percent supported construction of the dam, one-quarter was openly opposed and a further quarter either expressed no preference or would not answer, leaving only a minority in favour.[51]

King tried to minimize the significance of the results, observing that many of those who held back from supporting construction of the dam "own marsh of very little value," which had also been the situation the first time a vote was taken.[52] For his part, Roberts recognized that a further canvass would be required to get a majority in favour of the dam. He wrote to James Anderson, a Sackville-based member of the MMRA Advisory Committee, asking him to help recruit "a small group that might take an active part in securing the approval of a majority of owners." Roberts mentioned W.W. Fawcett, a prominent hay merchant in the region, as the type of person who should be able to "persuade those who are not greatly interested to favour the construction of a large aboideau."[53] Or, as King put it at a meeting of the provincial marshland commission: "The opposition to the aboideau has done a great deal of canvassing ... If those who were for the large aboideau did some canvassing there would be a substantial majority."[54]

And so the wheels were put in motion for a second canvass in the fall of 1953, and on this occasion nothing was left to chance. Fawcett wrote to Parker in late August, frustrated that the anti-dam people were active in making their position known, and arguing:

> A clear statement of the case from the opposite viewpoint is now due ... In the meantime, I am pleased to tell you a week ago eight leading supporters met privately, carefully went over the list of those who signed as neutral and each person present [was] assigned a member to visit and canvass. I do not know the results yet but hope we can change the percentage somewhat.[55]

The work of individuals such as Fawcett was reinforced by lobbying from the local newspaper, the *Sackville Tribune-Post*, which had supported construction of the dam from the start and came out swinging with an editorial aimed at "objectors who oppose the reclamation of land by this means." Echoing the language at the public meeting where the dam project had first been discussed, the newspaper scoffed at those who feared "that if such a project were carried out, a birthright will be taken away from the farmer that has been his since the time of his grandfather's grandfather." The *Tribune-Post* found that arguments in favour of tiding were "based on illogical and fallacious thinking," advanced by farmers who "fail to take advantage of the latest scientific developments."[56]

In response, Ern Estabrooks, who had attended all the meetings going back to Laurie Anderson's initial call to arms, published a letter to the editor objecting to how the farmers were being cast. Putting the matter in terms of competing forms of knowledge, Estabrooks noted how the newspaper "appears to consider the dissidents to be very short-sighted ... But the fallacy that they could run their farms better by following the advice of some arm-chair agronomists is a fallacy that has been proved only too often."[57]

In the end, Estabrooks's pleading was no match for the MMRA and its allies. When the results of the second canvass were announced in December, the number of opponents had not changed appreciably, but a significant number of those who had had no opinion were brought over to the "yes" side, so that 60 percent were now in favour.[58] The agency's experiment with democracy ended with the experts carrying the day over the farmers who wanted to continue their traditional practices.

The vote over damming the Tantramar might have been decided, but this did not entirely end the campaign for tiding in the Tantramar watershed. In his impassioned plea for continuing the practice, Estabrooks had feared that the impact of constructing the dam would be particularly felt in areas some distance from the river, where land would be permanently inaccessible to the tides, becoming "useless swamps that can never be reclaimed." This perspective was advanced shortly after the final canvass, when farmers from such locations lobbied the MMRA and its provincial partners to excavate ditches so that tiding could take place before the Tantramar Dam, now a foregone conclusion, was built.

In a brief to the New Brunswick minister of agriculture, the landowners of the Log Lake Marsh explained how they had 675 acres that were

> suitable for rehabilitation ... Away from the river and main tide ditches [these properties] did not receive as much silt and are lower in elevation. Being difficult to drain, many of these areas gradually reverted to cat tails and other peat forming vegetation ... With the correct number of tide

ditches, these lands will receive sufficient silt in two years to make them first class marsh. In view of the proposed construction of an aboideau [i.e., dam] at the mouth of the Tantramar River, it is urgent that these lands be tided at once.

The petitioners made it clear that opposition to the dam had not disappeared, even if the process of public consultation was over: "The tiding of these areas, previous to the construction of the aboideau, would remove almost all objections to its construction."[59]

In a technical document accompanying their brief, the Log Lake Marsh landowners explained what would be required to carry out this tiding exercise. In particular, they stressed that

to keep the silt on the lands, the outgoing water has to be slowed down. It is advisable to build a sluice on the end of the ditch [bringing in the silt)]... A clapper [gate] has to be devised, which does not obstruct the inflowing water, but has such a restriction that the velocity of the outgoing water is checked at all stages.[60]

This was a significant variation on the normal aboiteau gate, which was designed both to prevent incoming water and to encourage the rapid removal of water from inside the dykes. The landowners wanted the professionals who would assess their request to know that "tiding is not as simple as digging a ditch to allow the tide water to cover the land periodically, and then sitting back and watching the land build up. Constant attendance is required to keep the process going."[61]

If the goal of the Log Lake brief was to change the MMRA's position on tiding, it did not succeed, in spite of the marsh owners' efforts to show that this practice was based on careful study of how the tides could be most profitably controlled. In a sense, they were challenging the experts at Amherst to take the farmers' knowledge more seriously. Nevertheless, when the brief landed on Parker's desk, he immediately rejected any MMRA support, casting doubt on the willingness of "the owners of the marsh ... to contribute very much if anything toward the cost of tiding. I did not find anywhere in the brief that they have offered to defray any of the expenses." And even if the farmers were willing to pay their share of the costs, Parker doubted that the exercise "would be at all economical." Ending the conversation, he made it clear that "we do not have any desire to promote this matter of tiding."[62] On this subject, the MMRA consistently stood up against local knowledge, distancing itself from its earlier willingness to incorporate the farmers' practices into the reconstruction of the dykes and aboiteaux. This shift underscored how the MMRA had moved on to a new moment in its history with the construction of tidal dams.

Road Work

With the marsh owners' challenge to the Tantramar Dam over, the project moved slowly toward construction. The MMRA did not want to put a shovel in the ground before its first tidal dam, on the Shepody River, was completed in March 1956. But by the time the agency began exploring the costs of a project that had first been proposed years earlier, its financing had become entangled with construction of the Trans-Canada Highway.[63] New Brunswick needed to replace the bridge crossing the Tantramar River, which was not up to the highway's standards. As a result, Fredericton contacted Ottawa about piggybacking on its dam project, and the federal government agreed as long as the province paid for the additional costs that running a highway over the dam would entail. The two levels of government reached an agreement in 1957 under which the MMRA and the province shared the costs on the Tantramar Dam (including the highway), roughly on a 4:1 basis. As a result, when the dam was completed in January 1960 with a total budget of slightly over $1 million, the province had contributed roughly $200,000, half of which was reimbursed by Ottawa according to its cost-sharing arrangement with the provinces for construction of the Trans-Canada.[64] In this case, the MMRA project helped reduce the financial exposure of the province, and while the calculations would differ for the three dam projects that followed the one that blocked the Tantramar, all would feature a connection between dam and highway construction.

This merging of highway projects with MMRA construction was nothing new. Indeed, at the first meeting of the Advisory Committee in 1949, Parker explained that "he had been discussing the problem of combining an aboiteau [in the original sense of the term] with a main highway with the Nova Scotia Deputy Minister of Highways. The Deputy Minister had questioned the advisability of building such a structure at Hortonville [near Grand-Pre] on account of the small amount of marsh protected." By having the aboiteau also serve as a bridge, the costs for such a marsh might be justified. Anticipating the situation a decade later with regard to the Tantramar Dam, Parker looked forward to the time when "the Trans-Canada highways [created by legislation in 1949] would require new bridges up to their standards."[65]

The advantages of combining highway with dam construction took on particular significance as the MMRA moved ahead, shortly after the Tantramar project, with a dam that would block the Annapolis River in Nova Scotia. The idea of building such a structure went back to the 1940s, when it was promoted by the Marsh Owners' Association of Annapolis County, which wrote high-ranking officials in both Ottawa and Halifax, explaining that "the farmers of the area are much in need of the land which would be brought into production by the erection of a dam." After the MMRA's experience with the Tantramar Dam, it would never again seek the owners' perspective on its projects, but it would have met no opposition in this case, where a dam constructed near Annapolis Royal, where the river narrows, would protect roughly 4,300 acres of land, as opposed to reconstruction of the dykes, which would protect only 2,700. The Annapolis County marsh owners enthusiastically supported a project that would allow "more land to be protected from tidal waters than the present system of dyking would protect."[66]

In spite of this local support, early studies of the Annapolis project raised questions as to whether the significant construction costs could be justified solely on the basis of protecting marshland. Ben Russell, whom we met in Chapter 1, was an engineer with the Prairie Farm Rehabilitation Administration (PFRA) when he came east to assess the situation in the dykelands. Writing in 1943, he explained that a very large structure would be required to block the Annapolis River – as it turned out, one that was significantly larger than either the Shepody or Tantramar Dam. He observed that

> to determine the economics of such a plan, it would be necessary to design such a structure and compare it with the costs of running dykes and small aboideaux ... It is not possible to pass any off-hand judgment as to the merits of such a plan, but for what it is worth, taking into account the considerable difficulties of building a dam on a tidal stream and the size of the openings required, and the value of the lands which it would reclaim, the writer's opinion is that the cost would be prohibitive, unless such cost could be mostly charged up to some other purpose such as a [highway] crossing of the river or the development of power, or both.[67]

Russell's reservations aside, the building of a dam across the Annapolis River was one of the first such projects discussed by the MMRA. When the agency authorized a feasibility study in September 1949, the Treasury Board asked whether such a project could be completed "at a reasonable cost," a concern that was not expressed with regard to the comparable study for the Tantramar Dam.[68] Even before the release of the Annapolis study in early 1954, Parker tipped off A.W. Mackenzie, the Nova Scotia minister of agriculture, that the projected costs could exceed $4 million, or roughly twice the combined expenses for the Shepody and Tantramar projects.

These costs were linked to the sheer size of the dam, which ended up being 86 feet high and 2,500 feet long, as opposed to the Shepody Dam (the larger of the first two dams), which was 55 feet high and 1,645 feet long; or, to put it another way, the headpond created by the Annapolis River Dam would have five times the capacity of the one at Shepody.[69] Parker wrote to Mackenzie, pointing out that the Annapolis project "is going to be a costly effort ... Would you please advise me whether or not the Provincial Government considers the project worthy of discussion from the point of view of carrying out a joint program." If not, Parker was prepared to walk away from this dam, instead using the MMRA's funds for dykelands "which urgently require protection."[70]

In the case of the Tantramar Dam, the inclusion of the highway was the province's idea as it considered its contribution to construction of the Trans-Canada Highway. With regard to the Annapolis project, however, the MMRA was actively looking for a partner to absorb some of the costs, aside from the additional expenses involved with accommodating a highway. In that context, Parker made the provincial minister of agriculture aware of the fragility of the current bridge: "The Department of Highways have advised that the present bridge will carry present day traffic but that some time in the future it would quite likely have to be replaced."[71]

Disregarding this warning, the province dragged its feet, and the old bridge did indeed collapse only weeks before completion of the dam in 1960.[72] Over the seven years that Halifax dithered, the MMRA wondered whether it should proceed with the Annapolis project, in spite of the significant amount of dykeland that might be protected. At a high-level meeting in early 1954 to take stock of the MMRA's projects to date, when the Annapolis Dam came up for discussion, the federal deputy minister of agriculture observed that "it is understood that the Annapolis Dam could not be considered as strictly a marsh protection project." Parker then jumped in, confirming that "it would have to be a combined project."[73] Even when the project made its way to the Advisory Committee later in the year, Parker remained skeptical. Most of the members expressed unqualified support for the dam, the exception being New Brunswick's J.A. Roberts, who observed that "in his own mind he had turned down the dam project on the basis that the extra 1600 acres protected was difficult to justify at an extra cost of $1.5 million."[74]

The committee voted unanimously in favour of the project, but only a few weeks later, Parker (who was a non-voting member) broke ranks with his colleagues, something he had not done before and would not do again. He wrote the deputy minister:

> I do not feel that I can concur in the recommendation of the Advisory Committee ... This recommendation has caused me to give considerable thought to the Committee's reasoning. The project itself is one which I have had in mind since 1949 and during the early stages of investigation, it appeared feasible to me. However, in view of the difficulties which would be encountered during construction, which in turn increases the cost of construction, I consider that the capital outlay is too great for the benefits which may be expected.

He went on to report that there was one other committee member (most likely Roberts) who "changed his [negative] vote to make the decision unanimous."[75]

Even though the Advisory Committee had given its approval, Ottawa was not about to proceed without the participation of the province. At one point in 1955, Robert Winters, the federal minister of public works, told Parker that if Halifax was not about to make an appropriate contribution, "the proposed project should not be considered further."[76] It was only in June 1957, shortly after the Tantramar Dam cost-sharing agreement had been reached, that an agreement was finally signed that divided the costs between the two levels of government, with Ottawa paying roughly two-thirds of the cost for a project whose initial price tag had come down dramatically, to $2.5 million.[77] Construction then began in 1958 and was completed in July 1960, made possible by "round the clock work carried out on the dam since one span of the nearby provincial bridge collapsed about two weeks ago and completely disrupted traffic."[78] During the construction, a sign from the work site revealed how the project was as much about constructing a highway as it was about protecting drained marshland (see Figures 3.6 and 3.7).

The MMRA's insistence on incorporating a highway into the Annapolis project helped lower its costs, off-loading some of them onto the province, in the process creating a cost-benefit calculation that made sense given the additional land that would now be protected. Indeed, when the Advisory Committee considered the file, it dealt almost exclusively with such calculations, which had become the norm for all MMRA projects by the mid-1950s. In focusing on the finances, however, the committee ignored

Dam Projects

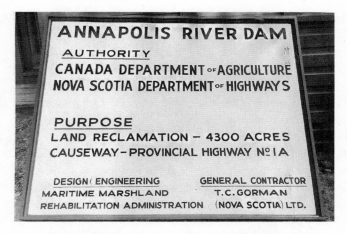

FIGURE 3.6 Annapolis River Dam work site. *NSDA. Reproduced with permission of Nova Scotia Department of Agriculture.*

FIGURE 3.7 Annapolis River Dam nearing completion. *O'Dell House Museum (Annapolis Royal, NS), Annapolis River Causeway Photos. Reproduced with permission of Annapolis Heritage Society.*

other matters that had been flagged by the feasibility study. For instance, the MMRA was slow off the mark in considering the potential impact of the Annapolis River Dam on fish stocks, even though the feasibility study had noted that "there is always a certain degree of doubt about the efficiency of any fishway installation or other means provided for fish to by-pass the rock fill to reach stream headwaters. The federal Department of Fisheries must give clearance on any proposal."[79]

This was not the first time that the MMRA had to deal with issues regarding the installation of a fishway, a device (such as a fish ladder) designed to allow fish to move easily in spite of the presence of an obstruction in a river. For instance, in the early years of the agency's operations, it received a request from farmers with land along the Gaspereau River, not far from Grand-Pré, asking that an aboiteau be built across the stream, in the process providing a highway bridge. Opposition soon emerged, however, from local fishers, including those who belonged to the Kings County Fish and Game Association, which feared that "the aboiteau would soon destroy the Gaspereau fish business, which has existed for many generations."[80] Due to this pressure, Parker concluded that "leaving the river open is more important to the Fishing Industry than the closing of it for purposes of developing new agricultural land."[81] He was prepared to look into the possibility of constructing a fishway, but concluded that his agency

> did not know of a suitable type of fishway which might be installed in an aboiteau which would dam the Gaspereau River. If we are able to design a fishway I would still be very cautious about erecting an aboiteau there as I understand the river is a means to the main spawning area for the gaspereaux in the entire Minas Basin area.[82]

In the end, the agency replaced the existing dykes and aboiteaux but did not obstruct the river.[83]

In contrast, when it came to assessing the impact of its tidal dams on fish passage, the MMRA seemed much less engaged. In the case of the Shepody Dam, no consideration was given to providing a fishway, thus creating a challenge for anadromous fish, species that needed to move freely between fresh and salt water. In the Shepody River, species such as salmon would begin their lives upstream from the dam, then make their way toward the Bay of Fundy before returning – again past the dam – to spawn. R.L. Butler, a federal fisheries officer in the region, observed in 1958 that such fish made their way past the dam only if "the gate operator opened the gates every day at the right time of tide to allow the fish past the dam to fresh water."[84] Butler's colleague, the engineer B.W. Hamer, noted that his

> observations at the Shepody aboideau [meaning dam] over several years have suggested delay to salmon attempting to enter the river in their autumn spawning run, and that perhaps alewives can also be affected seriously.... It appears that fewer [salmon] are entering than would if the aboideau did not obstruct them and that many are distracted from the gates to the basin next to the dam.[85]

In the case of the Annapolis dam, the inclusion of a fishway was first discussed in 1958, long after the project had been approved and roughly as construction was starting up, at a meeting in Amherst of MMRA and fisheries officials. As Parker put it, his staff was at the meeting "to go all out to work out satisfactory fish facilities with Department officials," and, toward that end, part of the meeting took place at the "MMRA's hydraulics lab on the outskirts of Amherst where a 1:120 scale model of the Annapolis Dam was set up."[86] As for the substance of the meeting, the transcript suggests a collegial affair during which the dimensions of the fishway were agreed on without very much difficulty, and it would appear that the opening did its job. In 1965, Dr. R.R. Logie, a senior fisheries official, wrote in response to concerns from an association of salmon fishers: "There is a permanent opening in the aboideau which is intended to serve as a fishway. We investigated the passage of fish for several years through this opening and could detect no holdup [in the movement of salmon]."[87]

But while Logie felt confident about the functioning of the fishway, he was less so about another situation, noting, five years after the closing of the Annapolis River, how "the aboideau slows down the clearing of pollution in that [upstream] portion of the river."[88] The question of how the construction of dams might affect the disposition of sewage had been raised as early as 1950, when the MMRA was considering construction of the Shepody and Tantramar Dams. A sanitary engineer with the New Brunswick government assured Parker that he need not worry about sewage that would be dumped into the reservoirs created behind the two dams because the communities there

> have no municipal sewerage systems and it is unlikely that their populations will increase to such an extent to warrant these public services. If the occasion should arise that sewage disposal in these areas is necessary, I am of the opinion that it could still be handled by the rivers and pass out with the fresh

water [from the reservoir]. For all practical purposes, I do not think that sewage disposal in these areas should be a factor in your present studies.[89]

With regard to the construction of a dam on the Annapolis River, a very different point of view was presented in 1954 in the feasibility study, which noted a potential problem because "raw sewage from towns and communities upstream from Annapolis Royal would be discharged into the stream and reservoir." Without the tide to flush out the offending material, it would simply stay in the headpond, leading the authors of the study to insist that "proper clearances be obtained from the Provincial Department of Health before the erection of a large structure is proceeded with and, if necessary, adequate provision made for sewage disposal."[90]

In spite of these warnings, and although there was a significant amount of talk, nothing of substance was done to deal with the potential problem prior to the closure of the river in 1960. The sewage issues were particularly significant for the town of Bridgetown, nearly twenty-five kilometres upstream from the dam but well within the section of the river that had previously been tidal. The town had been dumping its raw sewage into a creek that flowed into the Annapolis River. In particular, one of the town's largest employers, the Acadian Distillery, which had opened in 1957, was dumping its "principal waste material, spent grain, directly to the river."[91] With the dam closed, the town was faced

> with an acute, dangerous, and unsanitary condition. Having been denied the natural flow of the tides which were always of great velocity in two directions, the sewage now remains in the brook and the flow of water is insufficient to oxidize the sewage to public health standards ... The stench is causing a great deal of discomfort to the citizens of the town of Bridgetown.[92]

In response, the MMRA disclaimed responsibility for the sewage problem, Parker noting that "the discharge into an open creek ... [had been] an unsatisfactory condition prior to closure of the river." In any event, in the agency's agreement with Nova Scotia, "no specific reference is made to [its] responsibility for upriver works that might be affected by the construction of the dam." After months of wrangling, Parker finally agreed that the MMRA would assume two-thirds of the cost, up to a maximum of $25,000, to modify the sewer outlets, but even in buying some goodwill, he still did not see the situation as his agency's problem and reiterated his view that the town's position was "inaccurate as to the details of the situation."[93]

That these actions did not resolve the problem was made clear in a study of water samples taken in summer 1961, which showed high levels of pollution at a sampling station just downstream from Bridgetown. The report explained how "the scouring action [much reduced by the causeway] does not appear to be sufficient to prevent accumulation of sludge banks composed of spent grain [from the distillery]. In time these banks will become physically objectionable through the formation of floating sludge masses and offensive odours."[94] Four years later, as Logie noted, the situation had not changed.

In creating a new landscape, the MMRA's actions in the Annapolis watershed became the concern of individuals without any connection to marshland farming, individuals whose voices had been absent from discussions of either the reconstruction of the dykes and aboiteaux or the construction of the smaller dams across the Shepody and Tantramar Rivers. In 1957, at about the time the contracts for the Annapolis River Dam were being signed, the MMRA received a letter from C.E. Daniels and Son, a firm based in Annapolis Royal that sold recreational vehicles (its letterhead invited clients to "Live and Roam in a Rolling Home"). The letter supported efforts to "reclaim all the major and minor marshlands lying [upstream from the dam] on either side of the Annapolis River." But Daniels and Son was equally enthusiastic about how the area would "benefit from the maintained constant water" that would create "improved water frontage for all properties adjacent to the river and basin which in turn would result in increasing tourist attraction and the building of summer cottages in the area. In fact, far-reaching benefit would no doubt be apparent in many ways not previously mentioned."[95]

In spite of the pollution concerns that had already been expressed, Daniels and Son was looking forward to the creation of a freshwater reservoir that might attract tourists, and a similar perspective on the new landscape was expressed by the mayor of Bridgetown, who wrote to Parker asking that the MMRA keep the water level as high as possible for aesthetic reasons. As Mayor H.H. Smofsky explained shortly before closure of the river: "It is our opinion that the natural beauty of the Annapolis River at high tide constitutes one of the Valley's greatest assets and tourist attractions, taking precedence over the reclaiming of extensive marshland acreage." Smofsky was not objecting to the creation of a freshwater headpond, but he was rejecting the idea that the dam was only about dykeland.[96] Parker could try to wash his hands of responsibility for anything unrelated to protecting the land from the tides, but as the MMRA became involved in projects that combined various goals, such as highway construction, it was no longer involved exclusively in rehabilitation of dykeland, and individuals who had nothing to do with farming became part of the conversation.

Petitcodiac

The completion in 1960 of the Tantramar and Annapolis dams coincided with the virtual end of the MMRA's reconstruction program. As we saw in the previous chapter, with few projects left on the drawing board, the agency's Advisory Committee had no reason to meet between September 1960 and July 1964, and by the time it reconvened, Parker had moved on to a position in Ottawa and the MMRA had become part of the Department of Forestry, where it made its expertise available to the Agricultural Rehabilitation and Development Administration (ARDA) in regard to soil and water issues in Atlantic Canada, many of which had nothing to do with dykelands.

In spite of the appearance that the MMRA's work as originally defined was done, the agency spent nearly $15 million after the close of the 1960–61 fiscal year, or roughly 45 percent of total expenditures over its history; and the two largest projects over its last decade involved the construction of two further tidal dams, across the Petitcodiac and Avon Rivers. What distinguished these projects from earlier ones was just how little dykeland that was not already protected would be shielded from the tides. While there had been considerable debate over whether a dam across the Annapolis River was worth the investment when 1,600 acres of "new" dykeland would be protected, that figure fell to only 347 acres in the case of the Petitcodiac project and a mere 155 acres for the one across the Avon.

In other words, in both cases multi-million-dollar projects were essentially designed to protect dykeland that could just as easily have been secured by the less expensive reconstruction of the dykes and aboiteaux. No cost-benefit analysis was done with regard to the Petitcodiac project, even though such calculations had been the norm since the early 1950s. And when the numbers for the Avon project were crunched, it was clear that protection of the pertinent dykeland through reconstruction of existing structures was significantly more cost-effective than through construction of a dam.[97]

As in the case of the Annapolis dam, the final two projects were ultimately justified by the transportation infrastructure that could be constructed in the process. But while the Annapolis project had been initially pitched by local marsh owners, transportation concerns rather than dykeland protection were paramount from the start of the process of building the MMRA's last two dams. This was particularly clear in the case of the one across the Petitcodiac River, which was first discussed in the late 1950s in the context of the need for a structure that would permit easy passage between Moncton and its growing suburbs across the river.

There was already an aging bridge across the Petitcodiac, but voices emerged in support of building another link in light of Moncton's rapid growth during the 1950s, which led one writer to refer to it in 1957 as "Booming, Bustling Moncton." The city's population grew by 60 percent during the decade, fuelled by its role as a major transportation hub. Its leading employer, the Canadian National Railway, had a sprawling locomotive repair facility as well as its headquarters for Atlantic Canada in Moncton, where its workforce peaked at nearly 6,000 during the 1950s, earning over $20 million in wages. Maritime Central Airways, which operated out of Moncton, was "carrying more weight in freight than any other commercial Canadian airline." Major firms such as Chrysler, General Motors, and Westinghouse had their Atlantic headquarters in Moncton, where "most of them occupy brand new buildings on the widest street in town, known locally as 'The Golden Mile.'"[98] This growth led urban planner Harold Spence-Sales to claim that Moncton was "the fastest growing city in the Nation."[99]

Spence-Sales, a McGill University professor who founded the first planning program in Canada, was active in promoting master plans for a number of Canadian cities. He championed the construction of "circular routes" around urban centres that would "form a continuous orbital road that provides access between satellite aggregates and passes through open country."[100] In his Moncton study, carried out in conjunction with the city's housing committee, the ring road was possible only by building a new span across the Petitcodiac. This link would connect the city with the burgeoning suburbs across the river whose population grew by 170 percent during the 1950s.[101]

Although Spence-Sales claimed that construction of ring roads "should conform to topography and respect natural amenities," there was little evidence of such respect in his plans, which adhered to a single design regardless of the context.[102] In proposing a new road across the Petitcodiac, Spence-Sales failed to show the least interest in the fact that Moncton was located on a tidal river whose disruption would have consequences that at the very least needed to be examined. This was a far cry from the mētis exhibited by the MMRA in terms of the reconstruction of the dykes and aboiteaux, when the agency adjusted its practices according to the circumstances. Spence-Sales, however, was exhibiting classic high-modernist thinking. As James Scott has noted: "The lack of context and particularity is not an oversight; it is the necessary first premise of any large-scale planning exercise." With regard to such exercises that might affect the natural world, Scott pointed to the high modernists' narrow focus on a particular goal, in this case the crossing of the Petitcodiac, in the process ignoring other factors such as "the well-being of the community ... The clarity of the high-modernist optic is due to its resolute singularity. Its simplifying fiction is that, for any activity or process that comes under its scrutiny there is only one thing going on."[103]

In fact, the new span across the river might have taken the form of either a bridge (allowing the continued flow of the tides) or a causeway (blocking the river, as had the MMRA's earlier dams). Either option would have satisfied the transportation needs of Moncton, but with different implications in terms of both short-term matters connected with financing and long-term implications for the environment. The causeway option was aggressively pushed in early 1960 at a special meeting of Moncton City Council requested by the Moncton Board of Trade, at which Spence-Sales was in attendance. The head of the board, L.G. DesBrisay, opened the meeting by noting that he and the MMRA's Parker (also in attendance) had been "discussing the possibility of building a causeway across the Petitcodiac River." Acknowledging that there was another way to achieve the same end, DesBrisay mentioned "that the Provincial Government is interested in building a bridge."[104]

Parker then took the floor to promote the causeway option, which might provide a large project for an agency whose activities were drying up. His support of such a project in the early 1960s stood in stark contrast to his rejection of a proposal in the mid-1950s for the construction of a series of massive dams to block the entrance of the tides into both the Minas and Cumberland Basins, as well as the Petitcodiac and Memramcook Rivers.[105] On that occasion, even though the proposal was designed to create reservoirs that might be used to generate tidal power, Parker stayed resolutely focused on his agency's mission, questioning the appropriateness of spending millions of dollars when there would be only minimal returns in terms of the protection of dykeland. More specifically, with regard to the proposed expenditure of over $4 million to construct a dam at the mouth of the Petitcodiac River, Parker explained that it would have served no purpose in terms of dykeland protection as the job was already being done by the dykes and aboiteaux that the MMRA had reconstructed. He recognized that the dam might have brought some further land into production, but saw no reason to drain marsh for which "no request has been received."[106]

By the early 1960s, however, the context had changed entirely as the MMRA was trying to justify its existence now that the reconstruction of dykes and aboiteaux was largely complete. There was little dykeland left to protect, and the MMRA might have closed down, transferring its operations to the provinces as planned from the beginning. The Petitcodiac Causeway offered an opportunity for the agency to stay in business, and the project's promoters were enthusiastic about the involvement of the MMRA, with its expertise and financial support, which in the end came to roughly $800,000 out of a total budget of $3 million. As Janice Harvey has explained: "If the causeway were designed in such a way that water levels upstream could be controlled, the [MMRA] mandate to protect agricultural lands above Moncton could be met. For the municipal and provincial governments involved, this meant important federal dollars could be leveraged for the project."[107] In his presentation to city council, Parker indicated that his agency's involvement had little to do with marshland protection, a subject that he barely mentioned. Instead, he focused on the value of creating a freshwater reservoir upstream from the causeway, the headpond that came to be known to some as Lake Petitcodiac, which "would be of much more value to the City than the present bad features presented by the river," although those "bad features" were never enumerated.[108]

Parker recognized that "there would be problems [connected with such a causeway], but it was his feeling that the value would far outweigh the problems." He was questioned during his city council appearance by W.M. Steeves, the city engineer, about "the loss of the tidal bore," a distinctive natural occurrence on the Petitcodiac that had drawn the attention of visitors for over two centuries. In various contexts across the globe, where large tides (such as those in the Bay of Fundy) are concentrated as they enter a channel (such as the Petitcodiac River), a tidal bore can form, especially when the river bed is particularly shallow. As R.A.R. Tricker explains:

> A tidal bore consists of a body of water advancing up the river with the incoming tide and possessing a well-defined front separating it from the still or slowly ebbing water into which it advances ... As the advancing tide travels into the still water ahead of it, it gives it forward momentum and gathers it up, joining it to itself.[109]

Figure 3.8 Postcard of Petitcodiac tidal bore, 1906.

Figure 3.9 Surfers on the Petitcodiac, 1967. *Provincial Archives of New Brunswick, P115-12554.*

The river effectively *bores* into the ebbing tide. Regarding the Petitcodiac tidal bore, scientist F. Keith Dalton noted in 1951 how it began to take shape roughly twelve kilometres downstream from Moncton:

> At first it has a smooth wave front which may be seen, and also heard, breaking at the ends along both river banks as it proceeds. About eight miles further upstream, when the bore comes around the right-angle bend in the river at Moncton, it develops a bubbling front, averaging 3 to 3.5 ft. in height. At the highest tides and if supported by strong winds up the Bay of Fundy, the front may exceed a height of 5 ft. The bore advances about 8.5 mph, and to a distance of thirteen miles above Moncton.[110]

Dalton went on to describe how, a decade before the discussion of the causeway at Moncton City Council, visitors would come to the city expressly to view the bore from "a special park in Moncton where a platform is provided along the top of the river bank for the convenience of spectators."[111] The postcard from 1906 shown in Figure 3.8, one of many sent home by tourists to the region, spoke to the draw of the tidal bore.[112] But those visitors were unlikely to continue making the journey if an obstruction were constructed across the river at Moncton. The bore was possible only where the tide moving upstream washed up over the gentle downstream flow, but this flow would be totally transformed by a dam that would move the head of tide twenty kilometres downstream: when the bore came around the bend in the river, it would be clashing against the tides that were being pushed back by the nearby causeway.

Indeed, there was some indication of the likely fate of the bore even before the causeway was completed in 1968. In August 1967, four Boston-area teenagers came to Moncton to surf the Petitcodiac. Five hundred people waiting at the Bore Park were surprised to see the arrival of the surfers, who had been drawn by the prospect of catching a big wave as they went upstream on the tidal bore. The surfers were so excited that they even brought a friend with them to shoot film and take photos, one of which is shown in Figure 3.9. In the end, however, their experience did not live up to expectations, as the tidal bore had already been reduced as the "river was filling with mud because of construction of the causeway, which began the year before."[113] Writing in the 1980s, David K. Lynch noted how "bores, like many other phenomena on the earth, are vulnerable to the activities of man," and he pointed specifically to a tidal bore on the Colorado River that had been reduced by the construction of water control structures.[114] Parker seemed prepared to accept the same fate for the Petitcodiac tidal bore, explaining in 1960 "that this is one of the things that will naturally come about with the building of a causeway."[115]

Parker was similarly laconic when the city engineer asked about "the sewers which empty at the present time into the Petitcodiac River and which depend upon the flushing they receive from the rising and falling tides."[116] This was a matter that Parker would have known well from the problem that he and his agency were dealing with at the time along the Annapolis River near Bridgetown following the construction of that causeway, but the stakes were significantly higher for a large and growing city such as Moncton, with substantially more sewage requiring disposal.

At the time of Parker's appearance before Moncton City Council, the exact location of the causeway had not yet been determined, but both options on the table posed serious problems with regard to sewage. If the causeway were built downstream from the city, sewage would be trapped in the headpond, as at Bridgetown. But even if the dam were located upstream from Moncton (as would come to pass), the Petitcodiac's flushing action would be diminished by the presence of an obstruction forty kilometres from the river's mouth, leaving only twenty-one kilometres above the structure. Indeed, one of the defining features of the Petitcodiac project was that it was the one and only dam project located so far upstream. Going back to the Acadians, it had been understood that there were significant advantages to building structures such as aboiteaux (and later dams) close to the mouth of a stream to reduce the amount of silt that would be picked up by the tides and then deposited immediately downstream from the structure as the coarse elements settled out. In the case of the Petitcodiac, because the upstream location of the causeway would dredge up vast quantities of silt, the MMRA was running the risk that sewage would simply build up in mud flats just downstream from the causeway, never washing away with the now-diminished tides.

Indeed, by 1960, Parker already had first-hand evidence of how the location of a dam made a difference. The photographs taken by his agency (some of which appear earlier in this chapter) showed how the buildup of silt in the Shepody River (where only 13 percent of the river was below the dam) was much less than in the Tantramar River (where 34 percent of the river was downstream from the dam). He had to be aware that an even greater buildup of silt, enough to alter the shape of the river, was likely with regard to the Petitcodiac Causeway, where 65 percent of the river would be below the dam. But Parker did not seem too bothered by this prospect, saying nothing about the likely accumulation of silt and telling city council that "the problem of sewers is one that he has given considerable thought [and] is one which he feels can be worked out successfully." Based on such assurances, the meeting ended with a request for the city to have the federal government, by way of the MMRA, look into the possibility of building such a causeway, an initiative that Spence-Sales supported as it "tied in completely with his suggestions."[117]

The MMRA set out to conduct research, but long before its report was tabled in March 1961, it was drawing criticism, particularly from scientists connected with the federal Department of Fisheries who had caught wind of the Petitcodiac initiative. The involvement of such individuals who possessed their own form of expert knowledge would play a crucial role in the story of the Petitcodiac Causeway as it evolved over the subsequent half-century. Up to the early 1960s, the history of the MMRA had largely been shaped by the interaction between the agency's expert knowledge and the more traditional knowledge of the marsh owners. But now there were also other skilled professionals in the mix, providing their own assessments and in the process complicating the mobilization of such expert knowledge in the name of the state. While James Scott referred to the "state" as if it were a monolith, the MMRA's final dam projects along the Petitcodiac and Avon Rivers indicated that there were frequently significant differences in the advice provided by experts with different forms of training. Specifically, we will see how fisheries officials took into account the specifics of the situation, exhibiting their high-modernist local knowledge, something that the MMRA's engineers had long practiced but now seemed to be abandoning as they committed themselves to Spence-Sales's vision for the Petitcodiac project.[118]

Without getting into the significant impact that a dam might have on fish stocks in the river, the fisheries officers wondered why the MMRA, an agency committed to marshland protection, was even involved with the Petitcodiac scheme. R.R. Logie, the top fisheries official in Atlantic Canada, wrote to Parker in 1960: "This seems to be a rather odd function for your administration and we are wondering what land reclamation may be associated with the project."[119] Parker admitted that "the benefits pertaining to land reclamation are almost negligible as compared to other possible benefits," but pointed out, with a bit of an edge, that this was hardly "an odd function for this Administration as our staff has probably more experience constructing dams on tidal rivers than any group anywhere."[120] Logie's superior in Ottawa thought he understood why the MMRA director had replied as he had: "Mr Parker's irritation indicates that he is not entirely convinced that the prime purpose of his organization should be the assessment of highway bridges. On the other hand, he has no option but to carry out the instructions of his minister, using the small amount of reclaimed land as his entry into the field."[121] If this were the case, Parker was bending his expert knowledge to the will of his political master.

As for the impact of the causeway on the fish in the Petitcodiac, fisheries officers were quick to draw attention to the variety of species that needed to "ascend the river, includ[ing] salmon, shad, sturgeon, smelt, trout, stripped bass, alewives, eels, and tom cod." Focusing on salmon, R.L. Butler described the run as "quite extensive – estimates run in excess of 20,000 into this river and tributaries to spawn."[122] Thus, even though the MMRA's survey had still not been released in early 1961, J.L. Hart, the director of the biological Station in Saint Andrews, New Brunswick, confidently predicted that the causeway would "have a marked bearing on the hydrology of the area and thus alter, for better or worse, migration patterns of anadromous fish" such as salmon.[123]

Prior to the causeway, the river was roughly one mile wide, but the projected gate structure in the dam, which would be opened only intermittently to control the height of the headpond, was less than 10 percent of that width. In that context, the challenges for salmon stocks were underscored in a further study forwarded to the MMRA as it prepared its feasibility study. Dr. Paul Elson, a widely recognized authority on salmon stocks in the Petitcodiac and its tributaries, found that "a causeway at Moncton which incorporates a dam cannot but bring about profound changes in the upper estuary." He predicted that baby salmon (smolts) would end up being trapped "above the causeway [which] would almost surely result in heavy losses to predators, especially eels, and perhaps even more losses through heat stress with the approach of summer temperatures."[124]

Elson thought that some of the problems could be mitigated by constructing a fishway to facilitate movement around the proposed dam, but when the MMRA's feasibility study appeared in March 1961, there was relatively little direct reference to this issue. Instead, the agency claimed expertise in such matters from having recently constructed a fishway in connection with the Annapolis River Dam.[125] Logie returned to the fray, asking Parker about "the lack of liaison with our Department in the matter of the proposed Petitcodiac River causeway," and observing that it was too soon to have "positive evidence of the success of the slot and pipe fish passages in the Annapolis aboideau structure."[126] Parker responded in kind only weeks before leaving his post at Amherst, telling Logie that his agency had no obligation to consult with the Department of Fisheries because it was merely working as "consultants to the New Brunswick

Department of Highways." Sounding defensive, he did not feel required to discuss a report that was "semi-confidential," and the secrecy connected with the dossier extended to the determination by the New Brunswick government to keep the feasibility study out of the hands of the city of Moncton.[127]

The MMRA report, which shaped the developments that followed, was based on the assumption that a causeway should be built near Moncton. The report weighed the relative merits of various proposed sites, rejecting the city's preferred location, downstream from Moncton, because it would result in trapping the city's sewage, requiring an investment of $4 million (a greater expense than was needed for just the causeway) to extend sewage pipes through the causeway. In the process, the agency ignored the bridge option altogether, even though it might have avoided the problems that had already been identified with regard to sewage disposal, the buildup of silt, and the movement of fish.[128]

On the report's release, there was immediate pushback from those who wanted the possibility of a bridge to be given serious consideration. First in line was the New Brunswick Department of Public Works, whose report, only a few months after the MMRA's, underscored the fact that the new span, whatever form it took, had little to do with marshland protection: "*Traffic will be the prime function of any crossing.*"[129] The authors claimed that they were open to either a causeway or a bridge, but their evidence showed that

> a causeway system will require costly sewage collection and disposal facilities for all areas above the causeway site. Bridges ... will allow continued use of the river for direct sewage disposal. The location of bridge sites is therefore more flexible and bridges will in the total picture require less public investment for the same service than other alternatives.[130]

The connection between a causeway and sewage also figured prominently later in 1961, when the Town Planning Commission of Moncton explained how construction of a causeway would disrupt the town's

> natural sewer system ... The growth of Moncton for years to come would be held back by this sacrifice of our great tidal endowment. No foreseeable lake-shore development [along the headpond], costly to build and sustain, could ever replace in a money sense this unique natural phenomenon of cleansing tides which we and our great potential future city now have, FOR FREE.[131]

Much the same point was made two years later by Mayor Leonard Jones, who asserted that a bridge could be built for $1 million, or a small percentage of the total cost of a causeway and new sewage facilities. As he put it: "Right now – we need water and a crossing. Therefore let's get them both in the manner in which we can afford them – namely, a bridge on the western approach to the city."[132]

However logical the bridge option may have appeared to various parties, the causeway remained the crossing of choice for officials who were happy to take advantage of the MMRA's resources, both engineering and financial, which could be mobilized only if there was a possibility (however illusory) of protecting dykeland. The bridge idea never gained any traction, and the causeway project appeared on the agenda of the agency's Advisory Committee when it returned from its long hiatus in July 1964. But while the committee's discussion of the other tidal dams had gone into deep detail, and had included cost-benefit analyses, this one passed with relatively little scrutiny, underscoring the point that the MMRA's actions with regard to this dossier were not always based on analysis of the evidence at its disposal. Instead, it pressed ahead with a causeway that provided the illusion of protecting dykeland, in the process justifying its continued existence.

The Advisory Committee meeting began at 10 a.m., and first dealt with the reconstruction of dykes and aboiteaux at the Great Village Marsh Body in Nova Scotia, one of the few such projects still on the MMRA's agenda. As was the norm, the committee considered the agricultural value of the dykeland that would be protected as well as the pertinent cost-benefit ratio before approving the project. It then moved on to the Petitcodiac Causeway, with J.D. Conlon, the chief engineer, who chaired the meeting now that Parker had left, providing a history of the dossier, but without the sort of data that had been placed on the table for the much less significant reconstruction project earlier that morning. Committee members did not appear to ask any questions about this highly unusual project before approving it. The discussion was so superficial that the committee was able to adjourn before noon.[133] Much as Spence-Sales displayed a high-modernist willingness to apply a particular technique regardless of the context, so the MMRA assumed that its familiarity with tidal dam construction put it in a position to build such a structure independent of the circumstances.

With the agency formally on board, the Petitcodiac Causeway project moved ahead. A formal agreement between the MMRA and the New Brunswick government was signed in 1965, and excavation began in 1966 without any decision about provision for a fishway. Conlon was quoted as saying: "We're concerned about the fish problem, but we're not worried about it." By contrast, local sportsmen were "considerably less optimistic." Allen Kinnear, president of the New Brunswick Fish and Game Protective Association, observed "that a short bridge span incorporated into the causeway – meaning free flow of tide water through the structure – would be the best means of ensuring the continuation of the fish runs." Wilf Taylor, who wrote a weekly column on fishing for the *Moncton Daily Times*, also endorsed the bridge idea, adding that he couldn't "see how a fishway will work successfully as the water levels would not be at a stable level on the Bay of Fundy side of the causeway. As a result the fish would leave on or before the tide started to run out."[134]

The fishway issue remained unresolved until early 1967, when the Department of Fisheries approved installation of "a nineteen baffle vertical slot fishway" that would create a series of pools to permit the passage of fish, at least in theory.[135] A senior fisheries official confidently predicted the fishway's effectiveness, and specifically asserted that "the regular salmon run should be able to ascend past the dam without any major hold up or delay."[136] But almost immediately after closure of the Petitcodiac River in February 1968, it was clear that the fishway was not up to the task.[137]

In early 1969, R.L. Butler, the fisheries officer we met earlier, provided a detailed account of the impact of the first year of the causeway on fish stocks. In the months immediately following the closure, the salmon run was "far below the run of fish into this river and tributaries to spawn when compared to last year," and Butler observed much the same situation for other species, noting how there had been "a poor run of shad this spring [of 1968] into Petitcodiac river to spawn." He noted that when summer came, the water level in the reservoir was too low "for best operation of the Fishway ... It is possible that a few fish did enter, however many more were turned away and did not enter except when the tidal gates were raised in the dam." By the fall, "after the upstream height of water had been raised to desired level, a few salmon were checked through Fishway, considerably fewer fish than would ordinarily be entering the river."[138]

Butler had hoped that "over the years, fish will adjust to using this fishway," but he was unable to sustain that optimism, concluding that there would always be "some fish that this partial obstruction will turn away ... There is no doubt that this structure will have a detrimental effect on the Fishery of the area, Commercial and Sport."[139] Nor was there much optimism to be found in a column by Wilf Taylor, published only a few months after Butler's report, with the headline "Construction of the Petitcodiac Causeway Writes Finis to Migrating Fish Runs." Taylor quantified the

FIGURES 3.10–3.11 Petitcodiac River at Moncton, 1954 and 1967. Figure 3.11 shows the causeway beginning to block the river. Just to the right (downstream) from that construction, the darker-coloured silt has begun to reduce the width of the channel.

decline of fish stocks spawning in the Petitcodiac since the construction of the dam, and asked: "Would you consider this overall progress? I don't think you would even if you did stretch your imagination a great deal."[140]

Writing in 1970 with the benefit of another year of observation, G.T. Beaulieu, an engineer with the federal fisheries service, added to the gloom, describing in some detail the problems faced by "upstream migrants ... in their passage from the estuary into the freshwater tributaries to spawn. A regular flow of fresh water down the river and into the estuary is necessary to stimulate the salmon movement to start their upstream migration. Ideally this regular flow would be similar to the river discharges prior to the causeway construction." In practice, however, Beaulieu found that the outward flow of water when the gates were opened was too strong. Instead, there needed to be a "moderate river flow to attract fish from the estuary to the causeway and up through the fishway." But even when fish did successfully pass through the fishway, the opening of the gates sometimes caused them "to be swept back to the downstream side of the gates."[141]

In addition to the failure of the fishway, Butler chronicled another problem that the MMRA had downplayed in the years leading up to the obstruction of the river, namely, the buildup of silt downstream from the causeway. In 1963, there was a meeting of interested parties connected with the municipal and provincial governments (but not including the MMRA) to look for measures that might mitigate the impact of "tidal silt deposits [which] would cause the stream bank width to decrease downstream of a causeway and dam structure."[142] But such warnings were ignored, so that

photos taken before and during the construction process (see Figures 3.10 and 3.11) show a buildup downstream (to the right) at an early stage of the operation, effectively the situation encountered by the surfers. Only eight months after the closure of the river, Moncton City Council instructed the city engineer to watch over "the siltation buildup" and to arrange a meeting with relevant bureaucrats and elected officials "to conscientiously study the problems of siltation in the river."[143] Butler observed only a few months later that "the Petitcodiac river estuary is now rapidly filling up [in the] vicinity of Moncton city area with mud ... A definite narrower channel is now formed [in the] Moncton area of estuary and this trend will continue on downstream in the future."[144]

At the final meeting of the MMRA Advisory Committee in the fall of 1969, most of the discussion about the Petitcodiac project focused on the filling in of the river. J.A. Waugh, the chief engineer, was asked "if there was a problem with silting below the dam," to which he replied that "silting was not a problem insofar as the operation of the dam was concerned, but that there could be future navigation problems from the silt build-up resulting in a shallower and narrower channel."[145] The agency was able to move silt away from the dam by opening its gates from time to time, but it had nothing to suggest in terms of the rapidly changing shape of the river. Of course, it might have been prepared for this eventuality given its experience with the earlier tidal dams, where, according to one of its own officials, "the buildup of deposited material [had been] very rapid during the first twelve to eighteen months subsequent to dam closure."[146] But in this as in other contexts connected with the Petitcodiac Causeway, the MMRA seemed unconcerned about the implications of constructing an object in a tidal river closer to its headwaters than its mouth. Throwing caution to the wind, the agency created a situation that ultimately resulted in a decades-long battle, described at length in Chapter 4, to restore the Petitcodiac River from its status as Canada's "most endangered waterway."[147]

The Last Hurrah

The last major project that the MMRA saw to completion was the construction of a tidal causeway across Nova Scotia's Avon River. Much like the Petitcodiac Causeway, this one was dictated by transportation needs. Going back to the early 1960s, the Nova Scotia Department of Highways was interested in replacing "a highway bridge of inadequate size" as part of a "Valley Highway Route." Robert Ettinger, a trucker who frequently used the old bridge, told me how the bridge, not large enough so that two trucks could pass one another, would noticeably weaken whenever he stepped on the accelerator.[148] By 1963, the Department of Highways was joined by the Dominion Atlantic Railroad, which wanted to reroute its line that went through the town of Windsor so that it would run instead along the same span as the highway. For its part, the town, in addition to removing trains from its streets, stood to secure protection from flooding and creation of a freshwater reservoir behind the proposed dam.[149] The MMRA was called on "to determine the feasibility of constructing a tidal causeway across the [Avon] river," a task that its chief engineer, J.D. Conlon, embraced because of the agency's "close relationships with the contemplated causeways."[150]

As for the construction of a bridge, as plausible here as on the Petitcodiac, it was never discussed because the MMRA's participation once again needed to be justified by the protection of drained marshland. In the case of the Avon project, however, there was not even the lip service that had been paid to dykeland protection with regard to the one across the Petitcodiac. The documents justifying the project barely made reference to dykeland, which was not entirely surprising because the closure of the Avon would add only 155 acres of drained marsh to the more than 3,200 acres that had already been protected by the MMRA's investment of over $600,000 to rebuild the existing dykes and aboiteaux, mostly during the 1950s.

The protective structures were in good repair in the early 1960s, but this did not prevent farmers from viewing the project positively. Bob Wilson, a farmer with land along the Avon River, told me that "everybody that owned marshland all agreed the causeway was a good thing."[151] To be sure, the causeway stood to reduce maintenance costs upstream, but the MMRA's own analysis, tabled when the Advisory Committee met to discuss the file in June 1965, indicated that the cost-benefit ratio was significantly better for simply maintaining the dykes than for contributing nearly $2 million to construct a causeway that would add relatively little land to the farmers' holdings.[152] In addition to providing protection to dykeland farmers, the Avon project also stood to benefit the railway and the town, but those non-agricultural benefits were already factored into the MMRA's calculations, according to the formula created by Gordon Haase in the 1950s (discussed in Chapter 2). Based on past practice, the agency's calculations might have led to rejection of the project, but in this case – as with the Petitcodiac project – the agency was determined to employ its expertise, in constructing causeways if not in protecting dykeland. Even though the numbers pointed otherwise, the Advisory Committee approved the Avon project without batting an eye.[153]

Shortly after the meeting, G.J. Matte, Parker's successor as MMRA director, wrote to his minister to explain that there was really no rush to proceed with the dam. He observed that "the marshlands are presently well enough protected with dykes and aboiteaux," and recognized that a dam might be useful to replace the older structures "some time in the future." Matte claimed that the committee had recommended a contribution, but "they do not recommend actual construction now ... because there is no real urgency insofar as marshland protection is concerned. The only urgency is for the re-location of the highway and the railway."[154]

The MMRA had still not entered into a contract for the construction of the Avon Causeway nearly three years later, when the federal Treasury

Board proposed holding off even further until the MMRA had had an opportunity to respond to the 1967 report by Helen Buckley and Eva Tihanyi, which, as we have seen, cast doubts on the return that taxpayers were receiving from the agency's projects. The authors suggested that it might have been wiser to allow some of the drained dykeland to revert to salt marsh, since farmers were not using it adequately. More broadly, Buckley and Tihanyi found that marshland protection was now a "task completed," meaning that the Avon project could not be justified on that account; and there was the further consideration that by 1967 agreements had been signed with the provinces for the return of control over the protective structures, signalling the end of the agency.[155]

In that context, the Treasury Board was unwilling to authorize funds for the Avon project, as it "was concerned that if the construction of dykes, aboiteaux, and breakwaters is no longer economically feasible or even necessary, there is no point in continuing the MMRA program."[156] Within a month of what amounted to a cease-and-desist order, the agency provided its response to the report, making special reference to the infrastructure that had been created or protected as a result of the agency's actions: "Highways have been constructed or reconstructed, secondary roads developed, power lines and railways upgraded, housing constructed, sewage and water systems built."[157] With regard to the Avon project, these were precisely the selling points that the Treasury Board needed to hear, given the insignificant amount of dykeland that could be protected only through construction of a causeway. In short order, the Treasury Board committed funds so that the two levels of government could sign a contract in March 1968, with work beginning later in the year.

As it turned out, construction began just as initial reports were coming in about the impact of the Petitcodiac Causeway, a number of which cast doubt on the effectiveness of its fishway. No such structure was included in the Avon Causeway, however, even though studies carried out in the mid-1960s justified at least considering one. K.E.H. Smith, a biologist with the federal Department of Fisheries, wrote in 1965 that "the main river is totally obstructed by a series of power dams, pipeline diversions, and storage dams, some of which have been in existence since the early 1920s or before. No fishway facilities were provided in any of these structures." As a result, "the Avon River system presently has a very limited value to anadromous species ... Thus, construction of a causeway in the Windsor area would add little to the loss already experienced."[158]

Smith did not categorically reject the idea of a fishway, however. As he put it, although anadromous fish runs are very small, it may still be possible to maintain them, providing costs are not prohibitive. This may be possible by providing fish pass facilities similar to those on the Annapolis River aboideau [meaning dam], at very moderate cost. The fish pass there consists basically of a simple, rectangular, full depth opening, located adjacent to the regulating gates. This permits fish to pass at regular intervals during each tidal cycle. It has no regulating gates, but remains open at all times.[159]

To emphasize his point, Smith concluded that "installation of a fish pass [should] be seriously considered."[160] His perspective was reinforced only a week later by Neil MacEachern, another fisheries official, who, based on the same data, also concluded that the Annapolis model should be followed: "This would permit fish to pass at regular intervals during each tidal cycle and should be less expensive than the conventional type fishway."[161]

Three years passed after these mildly positive reports regarding a fishway for the Avon project, but when construction began in earnest in 1968, there was little interest in including a fish passage in the final design. Ultimately, the federal Department of Fisheries, which had the power to block any project that interrupted fish passage, came out against a fishway.[162] While recognizing, as had Smith in 1965, that there were modest catches for a number of species, a senior official noted that the department was no longer willing to install a gate for fish passage as at Annapolis "because of the intake of salt water. The Town of Windsor [was] very anxious to maintain a freshwater reservoir behind the causeway." Alternatives to the system installed on the Annapolis River were discussed, but the cost for such a fishway that might keep out the salt water was estimated at around $100,000, which was deemed prohibitive. Instead, the department suggested "that we explore the idea of using the freshwater reservoir for public trout fishing," evidently giving preference to the headpond (which would become known as Lake Pesaquid, becoming naturalized in the process), over the movement of fish through the causeway.[163]

While the decision was now final, there remained those in the region who continued to argue that there had been a sufficient fishery to justify the fish passage. For instance, Neil MacEachern, who had supported the idea of a fishway in 1965, was still bullish on the Avon River fish stocks, noting in late 1968, after construction had already begun, that "in the late 1960s, two local fishers alone had caught seventy salmon (mostly grilse) in just one season."[164] In a thesis that was heavily dependent on oral history interviews, Lisa Isaacman found evidence that supported MacEachern's views, noting one respondent who observed: "If you knew where to find them [salmon], there were all kinds of them." More generally, her interviewees asserted that "despite some declines (particularly in salmon and trout), fairly good anadromous runs persisted until the time of the causeway's construction."[165] Isaacman conceded that such memories may have been skewed by those who blamed the causeway for the loss of fish stocks, and so imagined "larger fish abundances" than had actually been the case. At the same time, she also wondered whether those in favour of the causeway "may have influenced the DFO's interpretation of the 1965 study results and led to an overestimated perception of the insignificance of anadromous runs."[166] In the end, all we know for sure is that a fishway might have been included for $100,000, a sum that was considered "prohibitive" in the context of a project that ultimately cost $3 million; or, to put the matter somewhat differently, it was one-third the roughly $300,000 that had to be found in order to route the town of Windsor's sewage so that it would not be trapped on the inland side of the causeway.[167]

While the Petitcodiac and Avon situations differed from one another in terms of how the fishway issue was handled, they were similar with regard to the buildup of silt downstream from the dams once the rivers were closed. In both cases, the tides had moved silt up and down the river twice daily for millennia. Bob Wilson, long before he was a farmer with land along the Avon, could remember playing as a child in a tidal river that had been his "playground ... In the summer time, we would play along the river, sliding down the mud bank, exploring what the tide brought in ... That all ended with the causeway."[168]

That mud now accumulated in vast quantities downstream from the causeway. Here, as with the Petitcodiac project, the MMRA left two-thirds of the river downstream from the obstruction. In the process, it departed from its long-standing practice, adhered to with regard to the earlier dams, of building the structures relatively close to the mouths of the river so as to minimize the quantity of silt that could be deposited. There was already a significant buildup of sediment at Moncton within months of the Petitcodiac River's closure in 1968, so it was entirely predictable that the same scenario would develop near Windsor on the Avon's closure in 1970, although there is no evidence that anyone was particularly concerned. Harry Thurston has described how "almost before the last stone was put in place, sediment began to accumulate to an alarming rate – five to fourteen centimetres per month. Within seven years, a four metre high island of silt formed on the seaward side of the causeway; and the effects were felt twenty kilometres downstream, where two metres of mud impaired navigation at Hantsport."[169]

Such developments could not temper the enthusiasm of the headline writer for the local newspaper, the *Hants Journal*, who announced, on completion of the causeway in June 1970, "ENGINEERS WIN – AVON RIVER CLOSED"; this enthusiasm was matched in a press release from the Nova Scotia government that trumpeted: "NEW CAUSEWAY FOR HOME OF THE WORLD'S HIGEST TIDE!"[170] This victory was particularly dramatic, given that the first attempt to close the river a week earlier had failed. Robert Ettinger was working on the job, operating a dump truck, and recalls how there was "just a little trickle of water [that passed through the rock fill], and she was gone. It was quite an experience. You couldn't believe the power of the water."[171] But things went better the second time, leading to excited descriptions, often couched in terms of a war that had been won. The Nova Scotia announcement proclaimed how "The mighty Fundy tides met their match," defeated "by an army of construction equipment." For its part, the *Hants Journal* wrote:

> Friday June 12 was an historic day for Windsor and marked the day man fought against nature, in the form of the Avon River tides, and won the battle ... As the morning wore on it appeared that practically every shopper in Town was dropping down to the river side to take a look ... There was a festive spirit ... As the hour to start dumping approached, the crowd increased until thousands were positioned at every vantage point on both sides of the river. The line up of earth-moving equipment on both ends of the Causeway, plus the thousands of spectators was reminiscent of an Armoured Brigade, preparing for a river crossing ... As the work went smoothly on, the interest of some of the spectators began to lag a bit, but was soon revived when spectators with binoculars reported that the "Bore" was starting up the river. From this point on the operators were racing against time and hundreds more spectators [some of whom are shown in Figure 3.12] arrived on the scene to see the final result ... The sight of the Avon River, with the mighty Bay of Fundy waters behind it, being held at bay by a man made dam was really awe inspiring ... At about 7:58 PM, the tide was fully in, took its customary rest period and then started to ebb. The thousands of spectators breathed sighs of relief. The battle was over. The Causeway was intact.[172]

WITH THE CLOSURE OF the Avon River, the MMRA's work was effectively done. Its agreement to transfer control of its projects to the provinces was scheduled for completion by the end of March 1970, so this really was the end of the line. In that context, the public celebration at Windsor, unprecedented in the more than twenty-year history of the agency, could be

FIGURE 3.12 Watching the closure of the Avon River, June 12, 1970. *Reproduced with permission of West Hants Historical Society.*

read as a tribute to an organization that had played a significant role in transforming the region's landscape. With the benefit of hindsight, however, celebration may not have been in order, particularly with regard to the final two dam projects, which departed from long-standing practices, both in terms of how projects were justified (ignoring cost-benefit assessments) and where they were constructed (ignoring the practice of locating structures close to the mouths of streams).

This behaviour was a long way from either the MMRA's careful attention to farmers' practices in devising its program for reconstructing the dykes and aboiteaux, or even its efforts to bring farmers into the process with regard to its earliest dams, which were devised in the late 1940s or 1950s. In the end, the MMRA and its provincial partners did everything they could to undermine the efforts of people like Laurie Anderson, who tried to hold up construction of the Tantramar Dam so that the practice of tiding might survive. Nevertheless, the MMRA recognized that the farmers' knowledge needed to be listened to, which is why the parties in favour of that dam kept holding referenda until they received the answer they wanted. However, the need to listen to others seemed to have been forgotten by the MMRA by the mid-1960s, and the final two projects had little to do with protecting marshland and almost everything to do with the hubris of experts.

Indeed, concern about the agency's going off script had been expressed in the early 1960s, just as it was starting to consider its final dam projects, when the Nova Scotia government convened a special meeting at Amherst to discuss "the future of the MMRA and the dangers of it becoming engaged in non-agricultural types of projects."[173] Reporting on this meeting, J.A. Roberts, New Brunswick's lead agricultural engineer, defended the MMRA, pointing to the benefits that had been reaped when its projects, such as the Tantramar Dam, not only protected drained marshland but also provided a highway bridge for the Trans-Canada Highway. What Roberts could not have anticipated, however, was that the agency would engage in projects where the benefit for marshland was negligible, taking on work whose primary purpose was outside its mandate, with consequences that it was unprepared to anticipate. The MMRA no longer existed beyond the spring of 1970, but its legacy in the region lives on, for good and for ill, as I write these lines a half-century later. This legacy provides the focus for the final chapter.

4
Legacies

In July 2013, two American surfers, JJ Wessels and Colin Whitbread, came to Moncton to break the world record for the longest distance surfed on a single wave. On arriving in town, they were looked at incredulously, the locals' body language saying: "Surfers? Here?"[1] Such a response, of course, made perfect sense, given the history of the Petitcodiac River, where the tidal bore, which the Americans hoped to ride, had been destroyed when the causeway was completed in 1968. In fact, as we saw in the previous chapter, the bore had already been compromised in 1967, while construction was still underway, when another group of Americans tried to surf the Petitcodiac. But much had changed between the two visits, a story that speaks to one of the legacies of the Maritime Marshland Rehabilitation Administration.

By and large, in the course of over twenty years of operations, the MMRA focused narrowly on the control of the tides, not always taking into account the impact of the structures it installed. Across the region, the agency transformed the landscape, providing farmers with large stretches of dykeland that they work to this day, and protecting – particularly through the construction of tidal dams – highways, railway lines, and a wide array of buildings, both public and private. In addition, however, there were various consequences of its actions that the MMRA ignored even when these were brought to its attention, either out of a lack of interest or based on the calculation that the advantages were greater than the potential costs.

The construction of the Petitcodiac Causeway was the most dramatic example of the long-term impact of the MMRA, as it became the focus of a debate that extended for over forty years until its gates were opened in 2010. The tides were now able to move up and down the river once more, allowing the return of the tidal bore, and with it the surfers. In addition, the river began to return to its pre-1968 state, with the erosion of silt that had built up and the return of fish stocks that had been destroyed. But before these changes could occur, there was a lengthy debate marked both by the sheer number of individuals who were involved, either directly or indirectly, and the diversity of the participants.

During its years in operation, the MMRA's engineers deployed their expert knowledge, sometimes working in concert with the marsh owners' local knowledge, other times ignoring what the farmers had to say. But now that the agency's work was done, there were questions as to what to do with the structures and landscapes it left behind, in the case of the Petitcodiac Causeway bringing into the equation the more than 100,000 people living in the river's watershed at the start of the 1970s.[2] Within this population, there were fisheries officials and environmental activists who adhered to an ecological approach to the marshlands, recognizing the connections between water, soil, humans, and other living creatures. As Mark McLaughlin has explained in terms of the rise of an environmental consciousness in postwar New Brunswick: "Ecologists demonstrated that humans could no longer simply conceive of the natural world as individual components that could be easily separated from one another."[3] This approach gave rise to a significant body of literature, including Rachel Carson's *Silent Spring*, which was nourished by her own observations about the challenges for salmon in New Brunswick's Miramichi River. Such works played a role in the emergence of the environmental movement and resulted, in terms of marshlands, in concrete actions such as the 1971 Ramsar Convention on Wetlands of International Importance, and, in the context of the Petitcodiac watershed, the campaign to secure the opening of the causeway's gates.[4]

While construction of the causeway engaged experts, some of whom were in the employ of the same state that had funded the MMRA, and drew the attention of a larger swath of the population that had embraced an ecocentric vision, there were also those who did not see destruction when they looked at the Petitcodiac. Indeed, the causeway created something new, and for individuals who built their lives along the headpond, what they called Lake Petitcodiac, it was important to push back and assert their own local knowledge of the environment they inhabited.

FIGURE 4.1 Surfers on the Petitcodiac, July 2013. *Courtesy of Trevor Gertridge.*

In the end, however, these residents lost the decades-long battle to save their lake, and with the opening of the causeway's gates, the surfers (depicted in Figure 4.1) rode the tidal bore for twenty-nine kilometres, their route ending at the causeway structure that remained in place. Halifax's Yassine Ouhilal, who reported on the adventure for the *Surfer's Journal,* noted how "as they came around the final bend, both banks of the river were lined with people as far as the eye could see – thousands of locals. Word had quickly spread that surfers were going to ride the Petitcodiac. With the river running through the town, out the people came, cheering and applauding."[5] Wessels and Whitbread easily broke the old record of eight kilometres, and they might have gone even further if not for the continued presence of the gate structure that blocked them from continuing another twenty kilometres, to what had been the head of tide before the causeway.[6]

The surfers' ride spoke to the legacy of the MMRA and the involvement of individuals with various forms of knowledge in shaping the landscapes it created. But while the Petitcodiac Causeway had the highest profile of any project in the agency's history, it was not alone in providing evidence, as we will see in this chapter, of both the fragility of this environment as well as the possibility for the further transformation, sometimes inadvertently and other times by design, of the landscapes, the various forms of nature, that the MMRA left behind.

The Return of Wetlands

Long before the MMRA turned its attention to building tidal dams, it focused exclusively on the reconstruction of the region's dykes and aboiteaux. In total, these protective structures shielded nearly 50,000 acres from the tides, and much of this area remains productive farmland to this day. However, it was apparent, long before the the agency closed its doors, that the supply of protected dykeland was outstripping the demand from farmers. This fact was driven home by a 1964 study that showed that only half of all marshland that might be used for agricultural purposes had been improved, meaning that it had been or was in the process of being drained. Since the MMRA had effectively done its job, the issue here was the failure (or inability) of farmers in parts of the region to carry out the work needed to drain their lands.[7]

In addition, there was also improved dykeland (land that had been drained) that was no longer being used for agricultural purposes, as was shown by C.I. Jackson and J.W. Maxwell, who examined land use in the Tantramar area of southeastern New Brunswick in the late 1960s. The study was prompted by "appeals by the Tantramar marshland owners for additional improvement work not already scheduled for implementation under the MMRA program. The provincial government decided that a thorough evaluation of future prospects for farming and other land uses was required before any additional inputs of public funds were committed to further works." The results of this evaluation were hardly encouraging, as the authors found that "the area of improved land is much larger than the area used regularly for crop and pasture production." In other words, after all of the work of the MMRA, there were large stretches in the Tantramar area where protected dykeland had not been improved, but even where improvement had taken place, there were "lands which, although once cultivated, are now under-utilized or idle."[8]

This conclusion was consistent with the findings of Helen Parson, whose study of "agricultural restructuring" in postwar Canada found that between 1951 and 1991 there were "extreme losses" in Nova Scotia and New Brunswick in terms of "the amount of farmland that went out of production ... These losses accompanied the widespread adoption of mechanisation in those decades. Many farmers on poorer land, unable to remain competitive, left farming." Across Canada, mechanization resulted in fewer farms and fewer farmers, but the decline in New Brunswick was the most pronounced as the province "lost 87% of its farms and 92.7% of its farm population."[9]

In a sense, the MMRA had done its job too well, approving nearly all the requests it had received from the marsh bodies and the provinces, in the process draining more land suitable for agriculture than was needed. For their part, Jackson and Maxwell suggested that this situation provided an opportunity to re-engineer the area's wetlands so that they might be welcoming to waterfowl that had lost their breeding grounds when first nature disappeared. After all, the name "Tantramar" was a corruption of the French *tintamarre,* meaning a loud racket, a description attached by the area's first French settlers to the sounds made by the large waterfowl population they would have encountered.[10] The authors concluded: "With much of the improved agricultural land now idle, some opportunities exist for improving the utilization of the area's resources by expanding the waterfowl habitat to include some of these idle lands." For centuries, "dyking and ditching practices have forced the present day populations of waterfowl into limited areas at the head of the marshes ... Up to the present, waterfowl and agriculture have been in competition with each other. There now exists an opportunity to end this competition."[11]

In particular, Jackson and Maxwell saw an opportunity due to the development during the 1960s of federal policy to reassess the use of farmland across Canada. With the goal of "rationalizing Canada's agriculture industry," Ottawa passed the Agricultural Rehabilitation and Development Act in 1961 to "encourage not only the consolidation of small farms into larger, more viable economic units but also the reversion of submarginal farms to 'natural' habitat."[12] In 1966, it tabled a National Wildlife Policy that called for the "purchase or long-term lease of large marshes which require management for greater productivity and public use." The policy recognized that "large marshes are important not only as breeding grounds, but also as areas where the birds may winter or rest during migration."[13]

This emphasis on migratory birds was consistent with the practices of the Canadian Wildlife Service (CWS), which would be a beneficiary of Ottawa's new policy in terms of repurposing agricultural land. Created in 1947, the CWS, according to its historian J.S. Burnett, was preoccupied with the enforcement of the Migratory Bird Convention, which was signed with the United States in 1916. As Burnett put it: "It is no exaggeration to say that ducks and geese dominated the ornithological agenda at CWS, at least during the first twenty years after 1947. Indeed, it has been suggested by some, and not always kindly, that the initials of the agency really stood for 'Canadian Waterfowl Service.'"[14]

In its early years, the focus on waterfowl was largely in terms of research, but with the new federal policy in the 1960s, the CWS turned to acquiring land where the waterfowl might thrive. National Wildlife Areas (NWAs) began to emerge near the end of the decade in southeastern New Brunswick and the neighbouring sections of Nova Scotia as "wildlife managers sought to return the more marginal dykeland to wetland habitat."[15] As Burnett explains: "Technically, the authority to designate lands as National Wildlife Areas would not be granted until the passage of the Canada Wildlife Act (1973) and the proclamation of Regulations in 1977. Nonetheless, the policy declaration (1966) was sufficient to launch CWS in the real estate business."[16]

Some of the NWAs were created out of the roughly 14,000 acres that the MMRA had chosen not to protect, allowing the land to revert to salt marsh. Most notably, this was the case with regard to creation of the John Lusby National Wildlife Area, along the Cumberland Basin, not far from Amherst, formed from 1,200 acres of unprotected marshland that the MMRA already found "out to tide" when it came into existence. In 1965, G.H. Watson, a biologist with the CWS, called for the acquisition of land at John Lusby, which was "the last remaining salt marsh in Cumberland County. It is a feeding area for thousands of geese during migration." Watson was one of a number of wildlife biologists who weighed in as to how dykeland might be given a new mission, in the process adding a new voice (and a new form of professional expertise) to the discussion. He explained that "the salt marsh is an important link in the chain of waterfowl habitat, provided by (the neighbouring) Missaquash and Amherst Point marshes, which are freshwater marshes. The loss of this salt marsh would mean that fewer birds would use the fresh marshes."[17]

A different dynamic was at work in the creation of the Tintamarre National Wildlife Area, where 3,800 acres had already been designated for protection by the time Jackson and Maxwell published their study in 1971. In this case, the legacy of the MMRA had to be reversed by re-engineering land that the agency had only recently protected through construction of new dykes and aboiteaux. When those structures were in bad shape during the 1940s, observers noted the return of waterfowl. As David Green put it: "As the salt water invaded the dykelands again, the waterfowl habitat was reborn. The birds began to move back in."[18] This situation was

FIGURE 4.2 Flooded hay field, near Sackville, NB, 1951. *Photo by Joe Boyer, Canadian Wildlife Service. Courtesy Al Smith.*

chronicled in the early 1950s by the ornithologist George F. ("Joe") Boyer, who was the service's only wildlife officer in Atlantic Canada during its early years.

In a series of photos from 1951, one of which is reproduced in Figure 4.2, Boyer showed hayfields that were now flooded, and he hoped that the decaying structures that had allowed the infiltration of the tides might be left alone, to encourage the return of salt marshes.[19] Indeed, he worked "for most of a year in a vain effort to forestall plans for draining a large section of the fertile wetland under the provisions of the Maritime Marshland Reclamation [sic] Administration."[20] In the end, however, the MMRA rebuilt the dykes and aboiteaux, and constructed a large drainage ditch in 1965.

W.T. Munro, a senior official with the CWS, observed that with the reconstruction of the protective structures, "the birds had to move back to the fringes once more," as "800 acres in the marsh body south of the village of Midgic [just north of Sackville] was drained and protected from the tides."[21] But it soon became clear that the land was not required by local farmers, and it was subsequently "acquired by the CWS a generation later and reflooded as part of the Tintamarre National Wildlife Area."[22] As part of the process, the MMRA's large drainage ditch was fitted with a water control structure that created a freshwater impoundment. As Green put it: "The process is similar to creating dykelands in the first place, but rather than building dykes to keep the water out, dykes are being built to keep the water in."²³

Most of the land that was assembled to create the Tintamarre NWA was owned by farmers who had abandoned their lands or were using it minimally. However, there were also properties that had indeed been rehabilitated by the MMRA and maintained by the farmers, but now needed to be reshaped once again if the land were to attract waterfowl. Take, for instance, the property of Ralph Oulton, which was described by William Whitman, another wildlife biologist with the CWS, as including twenty acres of "agricultural land composed of recently plowed land that was formerly classed as marsh by the owner." Because of the land's value, Whitman proposed paying $50 per acre, a price that had not been seen since the golden age of hay production, and urged that the CWS act quickly to buy up such properties "to prevent land prices from jumping, as they did in this case."²⁴ In the end, the CWS was forced to increase its initial estimate for acquisition of a single 3,800-acre section for the new wildlife area from $50,000 to $150,000 to preclude the marsh from being "endangered by drainage projects."²⁵ With the acquisition of this property, Joe Boyer's vision for the area was achieved, but only after a legacy of the MMRA's actions had been reversed.

Similarly, in creating the Shepody National Wildlife Area, also in southeastern New Brunswick, the CWS acquired land, some of which had been part of the New Horton Marsh Body. In approving the investment of $155,000 at New Horton in 1955, the MMRA's Advisory Committee called for the construction of a new dyke and two large aboiteaux because "all marsh [was] out to tide ... [There were] no existing protective works of any value." The committee was advised that "the owners were definitely interested in the reclamation," and yet only a decade later the land was not being farmed, so it was transformed into a home for waterfowl.²⁶ As the Shepody NWA Management Plan put it, this area had seen "significant manipulations of the habitat ... Wetlands have been cut off from the sea by dykes and then ditched and drained for agriculture, but water once again covers these long-abandoned agricultural dykelands through collaboration with Ducks Unlimited Canada."²⁷

The partnership of the Canadian Wildlife Service with Ducks Unlimited Canada (DUC) was crucial to the reversal of the MMRA legacy. Going back to the 1930s, DUC had been in the business of protecting wildfowl habitat, particularly Canadian breeding grounds for ducks whose numbers had declined, causing concern for American hunters who from the start were the organization's primary supporters.²⁸ At first, DUC focused on breeding grounds on the Canadian Prairies, but during the postwar period its activities spread, reaching the marshlands of New Brunswick and Nova Scotia. In the process, there was a recognition "of ecological connections – flowing water, wind, insects, seeds, or migratory animals – that regularly transcend property boundaries."[29] This was the same sort of ecological thinking that would be important in challenging the Petitcodiac Causeway. As Tina Loo explains, DUC "was in the business of re-engineering imperfect landscapes – usually ones that had been rendered that way by earlier attempts to drain them for agricultural use ... This was accomplished by building dams on the existing drainage channels to prevent water from flowing away, and by constructing dykes along the marsh to prevent flooding of adjacent fields."[30]

More specifically, in the case of the drained marsh that was being returned to wetland, Ducks Unlimited Canada was involved, in collaboration with the CWS, in installing water control devices that regulated the movement of fresh water from the dykeland that the MMRA had drained by way of aboiteaux. The new devices would prevent salt water from entering the marsh, but would also restrict the removal of fresh water, in a sense operating like the gates on the agency's dams that were occasionally opened to draw down the headponds if they were too high. These freshwater impoundments in areas such as the newly created Tintamarre NWA permitted the growth of vegetation that was fed on by invertebrates (particularly insects such as beetles), which in turn attracted breeding waterfowl.[31] However, these impoundments were still keeping the tides off the marsh, and in this sense were not returning it to first nature, reflecting an ongoing prejudice against the saltwater marsh as a productive environment. More recently, DUC has recognized the value of salt marshes and, with its partners, is "beginning to restore some of the original salt marshes to their more natural state ... Not only do these dynamic ecosystems buffer against storm surges, they are also among the most biologically productive habitats on Earth."[32]

In the National Wildlife Areas near the New Brunswick–Nova Scotia border, the second nature landscape that had been protected by the MMRA was transformed by the CWS and Ducks Unlimited Canada. The transition of this land from one function to another was personified through the role played by John Waugh, whom we met in the previous chapter. Waugh, a long-time MMRA employee, was the agency's chief engineer during its final days, before moving on to Ducks Unlimited as its Maritimes manager in 1971 as it started up operations in the region. The irony was not lost on Waugh, "who often point[ed] out that he was building freshwater marshes for ducks in the same places where he had once been in charge of draining salt marshes to produce farmland." He also observed that "he had spent so much of his lifetime draining marshes (regretfully, since he was a keen duck hunter), that he certainly knew where they were and how to reflood them!"[33] In this context, Al Smith, a biologist involved with the acquisition of land by the CWS, recalled how researchers studying the region in the late 1960s and early 1970s were "relieved that a use had been found for the land that the MMRA had protected."[34] In total, the NWAs on both sides of the New Brunswick–Nova Scotia border now protect nearly 10,000 acres of wetland, large parts of which had once been drained by the MMRA for the use of farmers.[35]

(Lake) Petitcodiac

The intervention of the Canadian Wildlife Service and its Ducks Unlimited partners led to the reversal of the legacy of the MMRA in ways that went largely unnoticed, even in the agency's own archival record. By contrast, efforts to undo two of its tidal dam projects attracted significant attention, some from individuals living far beyond the region, because of the visible impact of these structures on the landscape. In total, about 40 percent of the 80,000 acres that were protected from the tides by the MMRA were shielded through the construction of five tidal dams, but it was the final two projects, initiated as the agency was starting to wind down, that generated the bulk of attention, both at the time of their construction and in the half-century that followed.

The Avon and Petitcodiac causeways stand apart from those across the Shepody, Tantramar, and Annapolis Rivers as both were constructed largely to facilitate transportation projects, neither having much to do with protecting marshland. Moreover, the two were linked by their installation in tidal rivers, at some distance from the rivers' mouths, leading to the range of environmental problems discussed in the previous chapter. The project that generated the most significant controversy, ultimately resulting in the most ambitious solution, pertained to the Petitcodiac Causeway, which attracted significant media attention that engaged the region's population. According to Dr. Victor McLaughlin, the community was divided "along geographic lines, economic lines, language lines, and academic lines."[36]

McLaughlin, a highly regarded physician from Riverview, the suburb that benefited from construction of the new link across the river, was well positioned to assess the extent of public discord over the causeway.[37] He formed part of a group, the Lake Petitcodiac Preservation Association

FIGURE 4.3 Aerial photo, Lake Petitcodiac, 1995. Looking upstream, the silt-filled water below the causeway contrasts with the fresh water in the headpond above the structure. *PANB, RS 748/5270-P/ 0-95-32.*

(LAPPA), that strenuously challenged efforts to open or remove the five causeway gates, whose closure had created the headpond (shown in Figure 4.3) along which he and his neighbours had built their lives since the 1970s, enjoying the legacy of an MMRA-led project.

As McLaughlin tells the story, he and his wife moved to Moncton in 1957 determined to build their dream house there, something that could not be considered "ticky-tacky."[38] Toward that end, in 1964 they acquired land across the Petitcodiac and slightly upstream from Moncton. They could hardly have imagined that after completion of the causeway in 1968 they would find themselves living along a freshwater reservoir, what the McLaughlins and their neighbours always referred to as a "lake," in the process naturalizing this artificial body.[39] They constructed their home in 1972, with

> one feature ... which is rather unique in these modern times. There is no dishwasher. By design, both our dining room and the kitchen overlook the

lake. My wife and I wash the dishes together and we refer to it as our moment of togetherness. We talk about the day that is past and we plan the days that we hope will come. Frequently, the sun is setting while we wash the dishes and it is beauty beyond words and a tranquilizer that will never be matched by the pharmacy.[40]

From those windows, the McLaughlins watched their neighbours taking advantage of the headpond: "We enjoy the pleasure boats that wave as they go by. We enjoy the unknown friends in canoes who pass and raise a paddle in salute. We enjoy the windsurfers too occupied to see us. Above all, we enjoy the simple beauty of the lake, the beauty that goes with reflections on still water."[41] And Dr. McLaughlin's words were echoed by a couple that came to his neigbourhood much later. As Dorothy and Don Murray put it in 1997:

> Three years ago my husband and I bought a parcel of land in upper Coverdale for our retirement home. Last fall we built our home. We chose this site because of the quietness of the lake, the beauty and abundance of wildlife (Canada Geese, Mallard Ducks, eagles, loons, hawks, pheasants, etc.). Because this land was "waterfront" we paid more than we would have, had it been on the other side of the Coverdale road [away from the lake].[42]

Similarly, Ron and Darilyn Hill explained how "having this beautiful lake in our backyard has added value to our property, and removing the lake will definitely alter that value, considerably."[43]

The question of protecting their investment was important to individuals who had acquired property from land developers such as Bedford Buck, who told his own story in a letter to the New Brunswick minister of transportation, whose department was responsible for the causeway. Buck was writing to express his relief that Sheldon Lee and his colleagues had decided in 1992 that the gates of the causeway, already a matter of debate for over twenty years, would remain closed, assuring (or so it seemed) the continued presence of the lake. As Buck put it:

> I purchased expensive lake front property in 1978, long before there was any threat of the causeway gates being left opened. Since that time, I have developed what is known as "West Lake Subdivision" – where there are currently twenty-five occupied homes averaging more than $150,000 each ... Homes sell well here. Apart from three new constructions we started this month, there are no houses for sale. Houses in this price range are often

purchased for their aesthetic value. When a potential buyer looks at one of our new homes, it is impossible to overlook beautiful Lake Petitcodiac.[44]

Buck understood that life along the lake was dependent on one of the MMRA's legacies, the causeway, continuing to function as it had been constructed, with its five gates closed. However, the lake lifestyle was also dependent on the denial of nearly all of the 158 studies published over the last third of the twentieth century, almost from the day the causeway was completed, indicating that obstruction of the Petitcodiac River had resulted in the destruction of the fish stocks that had moved up and down the river for millennia.[45] As we saw in the previous chapter, fisheries officers warned about the potential dangers for fish in the river prior to the closure of the stream in 1968, but their expert knowledge was ignored, their political masters choosing to believe that a fishway would permit fish passage. In this way, Ottawa was able to issue the appropriate authorization to shut off the Petitcodiac.

With the causeway in place, the officers' worst fears came true, a situation laid out in stark detail by Conrad Bleakney, a fisheries officer based in Salisbury, where the head of tide had been located a further twenty kilometres upstream prior to 1968. Only a year earlier, he had observed: "I have never seen more salmon in the Petitcodiac River system since I started work with the department [of fisheries] in 1950, and the anglers had the best season ever in the area." And he had much the same to say about the smelt run, noting how "as usual these fish were very, very plentiful. Their numbers increase each season."[46]

Bleakney's 1967 report was echoed in a letter written decades later by Wayne Kaye, a fisher who recalled an experience on the Petitcodiac at roughly the same time:

> I don't really remember how many fish we hooked and released that day as we lost count, but it was well into the double figures ... On the way back to our vehicle, we commented that it had been a perfect day. The fishing had been great and we had not seen another fisherman. Our secret fishing hole was safe ... We fished that section many times from 1964 to 1968 when a marvelous salmon stream died. This fishing in many of these rivers and streams [that flowed into the Petitcodiac] was kept a guarded secret by many fishermen and in the long run, that was a mistake. That only helped the speedy construction of the causeway. Most people did not know and would not believe such stories, dismissing them as "fish stories." Only a handful who experienced it would believe it.[47]

Only months after the obstruction of the river, Bleakney reported that "for the first time in many years there were no salmon taken in this area by the anglers"; a year later, he described how "the annual smelt run was stopped at the causeway. Very few of these fish made it through."[48] Writing in 1979, he reflected on how he had worked for

> twenty-nine years ... to build up the Salmon stocks in the system by keeping the poaching to a minimum. We were also helped to a very great extent by a research program ... where hundreds of thousands of Fry, Parr, and Smolt [various stages in the life of young salmon] were planted in this River ... As a result of this program we were rewarded with thousands of Salmon in our Rivers before the Causeway. Since the Causeway our Salmon Population has steadily decreased to almost NIL.[49]

But the problems stemming from the causeway went beyond the destruction of fish stocks, as if that weren't enough. As had been anticipated, the deposit of vast amounts of silt downstream from the causeway resulted in the significant narrowing of the river. Gary Griffin, who was involved for over thirty years with efforts to open the causeway's gates, noted in 1979 how "tidal silting has transformed the mile-wide muddy river bottom," a situation confirmed by aerial photographs (such as those shown in Figures 4.4 and 4.5) that showed how the river was now no more than a narrow channel. In the process, the tidal bore all but disappeared, leading Bleakney to observe that by the end of the 1970s, it was "a mere ripple." The fisheries officer "watched and listened to tourists seeing this, the most famous tourist attraction in Eastern Canada, for the first time and heard them snicker. Yes, some even laughed out loud." The tidal bore was now commonly referred to as the "total bore."[50]

In 1979, as the evidence mounted, the New Brunswick Department of Transportation, which operated the causeway, hired the engineering firm ADI to examine the situation, with a particular focus on how the fishway and the gates might be adjusted to permit fish passage. The ADI report rejected the option of opening the gates, a solution that had been advocated earlier in the year by J.R. Semple, a biologist working for the federal Department of Fisheries and Oceans (DFO) in Halifax, who found it "to be the best means of assuring fish passage at the causeway."[51] The ADI engineers were not sold on this solution, expressing apprehension about what would happen to the accumulated silt if full tidal flow were restored, although they provided no evidence to support their concerns. Instead, they called for "modification of the gates to seal effectively in both directions,"

FIGURES 4.4–4.5 Petitcodiac River at Moncton, 1963 and 1976. *Aerial photos obtained from New Brunswick Department of Natural Resources.*

assuming that the problem was with the gates, which had intentionally been designed to leak so they would not "get silted up and become inoperable."[52] ADI's preferred course of action presumed that fish had avoided the fishway because they were attracted instead to the water leaking through the gates. This conclusion sidestepped the point made in numerous studies that "no fishway design currently exists to accommodate all indigenous species on their migration to traditional spawning and feeding grounds and permit the passage of juveniles and adults alike on their return journey to sea."[53]

Despite serious reservations about the ADI plan, Fredericton decided to go along with it, leading to an avalanche of letters from fishers who felt

let down. Typical was the letter from James Barron from Riverview, an "avid fisherman," who wrote to the provincial minister of transportation that he was

> concerned about the effect of the Moncton-Riverview causeway on the fish supply ... Because the fish can't get upstream, one cannot fish smelts at the bridge in Salisbury as once many people did. I know that small event does not overly concern you, but many people enjoy the sport and the "feast" after. Why can't this situation be rectified? Could a span of some sort be put in the causeway so the fish (not just salmon) could go upstream as Nature intended?[54]

Fishers such as Barron and others who shared his view also had a chance to express themselves when they dominated a public meeting held in the fall of 1980, the first of many such assemblies staged by individuals on all sides of the causeway issue. On this occasion, 250 people packed a hall, a newspaper report noting that "all but a few [were] clearly in favor of the installation of a bridge span."[55] According to a report on the meeting from Thomas Pettigrew, a biologist from the provincial Department of Natural Resources (which favoured opening the gates), the atmosphere in the room was set by the local MP, Gerry McCauley, who said that "he personally 'leaned towards blowing the causeway out of there' [or some statement to that effect]." Pettigrew observed that the Department of Transportation (which was opposed to opening the gates) did not even bother to send a representative, and he was not surprised when "little concrete information came out of the meeting."[56]

In fact, the only person on hand to respond to the calls for removing the gates was Daryl DeMerchant, an engineer from ADI. In response to such calls, DeMerchant said that "it would be a brave engineer who would guarantee what would happen then," and was greeted "with hoots of derision."[57] Pettigrew observed that DeMerchant "spoke briefly on the risks of removing the gates, but, at least in my opinion, failed to touch on what benefits were to be gained or lost." Near the end of the assembly, when DeMerchant was asked whether "he had learned anything from the meeting, his response was that nothing he had heard would alter his recommendations." Pettigrew found that "this type of response tended to cause many people to feel that they were wasting their time."[58]

If both scientific studies and popular support seemed to be on the side of opening the gates, then why did the government persist (for another thirty years) in keeping the gates closed? Some insight into this question was provided, shortly after release of the ADI report, by E.D. Gilchrist, an engineer who headed the Engineering Branch of the provincial Department of Agriculture and Rural Development. This was the same office that had worked closely with the MMRA as it reconstructed the dykes and aboiteaux across the province to drain farmland, largely in the 1950s. Gilchrist made reference to the "4000 feet of dykes that have been levelled since the causeway was constructed and would have to be rebuilt if the river again became tidal."[59] At the same time, he recognized that there were only eight commercial farms over the twenty kilometres stretching from the causeway to [the former head of tide at] Salisbury, an observation that underscored that the MMRA's actions with regard to the causeway project had had little to do with draining dykeland for farming.

Gilchrist went on to note how this area "is fast becoming built up on both sides of the river ... Many of the residents of the area of concern have moved there since 1968. If the river is changed back to a tidal river, the aesthetic effects may not be acceptable to them."[60] It is worth recalling Bedford Buck, the land developer we met earlier. He had made a significant investment along the headpond in 1978, only a year before the ADI report put the range of options on the table for public discussion, and was not likely to take kindly to any tampering with the gates that might undermine what he recognized was "a very risky business."[61] There already was a standoff between the weight of scientific evidence and the interests of the Lake Petitcodiac landowners.

EXPERT KNOWLEDGE CONCERNING the Petitcodiac was largely ignored by the Conservative government of Richard Hatfield, which led New Brunswick from 1970 to 1987, effectively the first two decades following the obstruction of the river. Nevertheless, in 1982, there was a brief glimpse of what the river could become again if the gates were opened. A report noted that "construction of a water main under the Petitcodiac River upstream of the dam required that the headpond stay drained during late spring and early summer by opening the gates at low tide and closing them on the incoming tide," effectively operating like an aboiteau. "This accidentally helped smolt fish passage downstream past the dam. The result was a phenomenal return of salmon in 1983 that has not been repeated." It is little wonder that this apparently successful (if accidental) experiment was not repeated, as it drained the headpond, destroying the "lake."[62] Oddly, the federal government continued to quixotically stock the river with salmon during the 1980s, spending roughly $3 million in an effort that was doomed to failure as long as the gates remained closed.[63]

The dynamics changed completely, however, with the election of Frank McKenna's Liberal government in late 1987, supported by individuals such as Gary Griffin, who belonged to a small group that had been lobbying since the early 1980s for the opening of the gates.[64] Active in the provincial Liberal Party in the Moncton area, Griffin wrote shortly after the election to the incoming minister of natural resources:

> Removal of the gates, to facilitate restoration of the Petitcodiac River, offers the government an opportunity to spend tax dollars and receive a return four to sixteen times the investment ... Short-sightedness caused the loss of one of the greatest attractions in our province; foresight would demonstrate

that we appreciate the value of our natural resources and are willing to correct past mistakes.[65]

Griffin's timing was excellent since the McKenna government appeared interested in various environmental initiatives at the start of its mandate, establishing a Round Table on Environment and Economy in 1988, and opening the causeway gates from April to June, and again from September to October, that same year. In the process, fish migration was encouraged during key parts of the season. A DFO report only a month after the spring opening noted that

> positive results have already become apparent. This includes the return of the smelt fishery near the village of Salisbury; shad caught near the village of Petitcodiac; and sea trout caught at various up-river areas. The downstream migration of Atlantic salmon smolt will not be subjected to delay and potential predation this year and therefore should result in some increased numbers of returning adults. Residents are also enthusiastic about the return of the tidal bore.[66]

Jack Powell, a long-time fisher along the Petitcodiac who had seen the decline of the river, "saw an immediate improvement in his catch. 'I got 1600 pounds in one pull in two hours – about 500 shad.'" Delighted by the opening of the gates, he explained that "the shad is good as long as the water is circulating. They won't come up when the gates are closed." As Janice Harvey put it, "it appeared some momentum was gathering for the permanent opening of the gates."[67]

But such success stories left the residents along the headpond and their allies unmoved. At the end of the 1988 season, John Betts, a Moncton city councillor who consistently opposed opening the gates, painted the experiment in the darkest tones, making reference to the silt that had been accumulating for twenty years and was now washing back into the headpond, without mentioning that full tidal flow (and not just this five-month experiment) would return the river to something approaching its pre-causeway state. Moreover, Betts asserted that the waters now flowing upstream above the causeway had been polluted by sewage dumped into the river downstream from the structure. Naturalizing the headpond, Betts feared that its "fragile ecology [would be] severely disrupted." Writing in such a manner, he presented the defenders of the causeway as environmentalists who were using their local knowledge to protect the reservoir that had now become a mix of mud and pollutants. As Betts put it, this resulted in a "decrease in the number of boaters and sailors. A navy regatta with one hundred entrants also had to be cancelled because of the gate openings."[68]

In spite of Betts's claims, tests carried out shortly after the gates were opened provided "no evidence that the pollution [in the headpond] is coming from the lower side of the Petitcodiac." In fact, a number of upstream sources of pollution were "identified and corrected" so that "the most recent tests show pollution levels dropping."[69] And in spite of significant evidence of increased fish stocks in the river after only one season, Betts made no reference to such inconvenient facts. In the end, he made the argument that would be made many times in the years ahead: "We presently have a lake, but we are not guaranteed a successful fishery even if the gates were opened. We do not want to gamble away a sure thing for something uncertain."[70]

But the weight of evidence didn't seem to matter, because the naysayers won the day, at least in the short term. After the five-month experiment in 1988, the gates were opened for only one month during each of the subsequent two years, and only during low tide to satisfy the concerns of those who wanted to protect their lake from the intrusion of water moving upstream, in the process restricting fish passage. The McKenna government then ended gate openings altogether for 1991, in advance of an election that fall, commissioning further studies so that it could make a final decision at a later date.

As the election loomed, Fredericton appeared uninterested in the conclusions of scientists such as John Ritter, the director of the Freshwater and Anadromous Division of DFO for the Fundy region, who explained that only "a free-flow system" would allow the return of fish stocks.[71] Instead, the McKenna government assessed the political costs of any decision. Interested parties sent hundreds of letters to the premier and his pertinent ministers in the months leading up to the 1991 election. For instance, Victor LeBlanc, a fisher downstream from the causeway who had long argued for the removal of the gates, wrote to the minister of transportation a week before the vote: "We the fishermen and concerned citizens are confident that the Liberal Party will react to this problem and foresee in the future that causeways are no solution. We know that the McKenna government has a difficult decision, but there seems to be only one thing to do and this is keep the Petitcodiac River causeway gates OPEN."[72]

At the same time, the government was also lobbied by members of the Petitcodiac Lake Preservation Committee (a forerunner of the Lake Petitcodiac Preservation Association). Edgar Mitton, its chair, wrote a series of letters in the lead-up to the election, reminding McKenna that the causeway's opponents talked "as if fish were the only inhabitants of the Petitcodiac river system." Instead, Mitton wanted to talk about how "the land which borders on the lake has provided nesting grounds for water fowl, birds of many different species, muskrats, etc.," an environment that would be threatened by opening the gates. "Mr. Premier, we implore you to break free from the gun barrel vision of those who see fish as the entire ecology."[73]

As the election approached, such lobbying appeared to be having an impact. J. Sydney Bird, who wrote numerous letters to McKenna in opposition to the causeway, took to task Hubert Seamans, Liberal MLA from Riverview, for appearing to side with the residents of his riding who lived along the lake. Bird observed that "at the present rate the river will only be a narrow channel within a year. As narrow as the minds of those who only wish to play on the lake and have no thought or care about the rest of the river system." As for Seamans, Bird explained that "a few hundred votes are worth more to [him] than a few thousand salmon ... It is well to remember that it is not a river system but a political system that has doomed the salmon, shad, and smelt."[74]

Seamans understood that there was anger against his government for having opened the gates during its mandate, but his efforts to return to his constituents' good graces failed, as he was defeated in the 1991 election, as was the Liberal MLA for the Petitcodiac riding that also bordered the headpond. The Liberals remained in power, but rumours quickly spread that with these defeats "the efforts to let fish up past the causeway would be abandoned."[75] For his part, Gary Griffin wrote to McKenna, asking whether "the politics of the present situation [have] made success [in regard to the Petitcodiac] an impossibility."[76]

Griffin and his colleagues remained hopeful, however, particularly following the release in May 1992 of the report of the Premier's Round Table on Environment and Economy, which had been established early in the McKenna government's first term, when it was trying to appear green. To develop its report, the Round Table held hearings across the province, including one in Moncton in late 1989 where Griffin explained how in the previous year, there had been "162 different people who used the lake, [but only] eighteen came back more than once. Everybody else came once. There's absolutely nothing to see or do on the lake – as far as fish life goes, it's dead. The only value the lake has is the aesthetic value for the homes that are along it."[77]

No one in favour of keeping the gates closed testified, with the result that when the Round Table recommended that the government "make a firm commitment to protect our lakes, rivers, watersheds and shorelines from degradation," it specifically proposed making the restoration of the Petitcodiac River one of a small number of demonstration projects. The Round Table wanted to highlight how "integrated environmental and economic planning can work effectively ... Restoring the river could serve as a model for other river rehabilitation throughout the province."[78]

In spite of this high-sounding talk, the New Brunswick government decided in late June, only a month after release of the Round Table report, to keep the gates closed, an action that appears, on the basis of cabinet documents, to have been based entirely on the fact that it was less expensive to maintain the status quo than to open the gates, which would have required the reconstruction of the MMRA protective structures that had either deteriorated or been removed over the previous quarter-century.[79] Nevertheless, it is difficult to ignore the fact, highlighted by the New Brunswick Department of Environment, that "a small group of people who own land on the lake was lobbying very hard to keep the gates closed."[80]

Given the context in which the decision was made, Griffin, writing with Julia Chadwick, expressed shock at how "the Government applied politics to the Round Table recommendations, completely negating the entire round table process. The government's decision was to ignore the environmental and economic benefits for the majority of residents, and let the Petitcodiac River and its unique ecosystem die."[81] More personally, Griffin expressed his sense of betrayal in a letter to McKenna:

> It was with total devastation that we listened to the media coverage on the future of the Petitcodiac River System. The news release portraying the decision [to close the gates permanently] is totally dishonest ... Your government did not have the courage to tell our communities the truth. Because of your previously expressed views, we did not expect a totally one-sided, anti-community, anti-environmental decision.[82]

By contrast, property owners along the headpond savoured their victory. For instance, Bedford Buck, the land developer, wrote to the minister of transportation: "I was relieved when I heard of your decision to keep the gates of the Petitcodiac River closed." Such relief was echoed in letters sent by various lakeside homeowners. For instance, Douglas Hamer congratulated the minister

> on his courageous stand ... Since I became married and raising a family, I have built two homes overlooking the now Lake Petitcodiac. My children

learned to skate on its frozen surface for the past fifteen years; we have cross country skied over it in the winter and boated in it and enjoyed its view for the almost twenty years we have lived next to it.[83]

For their part, Gary and Jane Sherrard wrote the minister that "as landowners (and we're not going to make apologies for our place of residence), we feel that you made the right decision." After describing how "our two children can go for a swim in an unpolluted freshwater tributary," they closed: "It is unfortunate the public does not believe that our wanting the lake to remain has absolutely nothing to do with depreciation in housing."[84]

THE UNEXPECTED DECISION to keep the gates closed totally changed the controversy, which had long been more about politics than the science. As one DFO scientist recognized, "the Causeway problem is very political (for and against)," and would require "a political solution."[85] McKenna may have calculated that there was only a small number of activists who really cared about the causeway, allowing him to side with the property owners along the headpond. But this decision backfired, as it mobilized wide swaths of the population. A memo prepared for cabinet in the fall of 1992 indicated that

> subsequent to the Cabinet decision to keep the gates closed, a number of departments received intense pressure to change the decision or at least reopen the issue ... Residents of southeast N.B. have created a broad based coalition representing the interests of fisheries, tourism, wildlife and the environment in support of opening the gates and have presented the government with a petition with more than 1000 signatures.[86]

In addition, a well-organized campaign resulted in hundreds of identical letters, in both English and French, being sent to Fredericton. The signatories expressed puzzlement over the government's rejection of the Round Table's recommendations and called for the reversal of "une décision qui permet la continuation de la destruction du système de la rivière Petitcodiac."[87]

As the anti-causeway activists stepped up their campaign, they mobilized segments of the population that had not been previously involved. The distribution of the form letter of protest in both languages reflected an effort to connect the Acadian community with this cause. While individual Acadians, such as the fisher Victor LeBlanc, had been involved with the opposition to the causeway since the early 1980s, there was now an effort to connect Acadians more broadly with a landscape that had originally been transformed by their ancestors, as celebrated in the film *Les aboiteaux* almost forty years earlier.[88] In the intervening decades, however, Acadians had not figured prominently in the affairs of the MMRA, nor were they particularly visible in the Petitcodiac Causeway affair until after the government's decision to keep the gates closed.

In that context, when Gary Griffin and his colleagues staged a Petitcodiac River Symposium in September 1992, it was attended by many of the usual suspects, but also on hand was Paul Surette, from the Centre d'études acadiennes at the Université de Moncton, who explained that "all Acadian culture is intimately tied to the tidal marshes," linking Acadians to this landscape, particularly as it was transformed from first to second nature prior to the deportation.[89] This sentiment was echoed and expanded on a few years later in a letter in French to the premier from Michel DesNeiges, who led the student group Écoversité, based at the Université de Moncton:

> In the past, the Petitcodiac River was closely connected to the survival and growth of the Acadian people ... The river has played a key role in the collective identity of Acadians. It speaks to Acadians' mythical past. Just as the tides have to persist to make their way up the Petitcodiac, so too have the Acadian people persevered across the highs and lows of their history.[90]

Couched in these terms, the fight to save the river became part of a larger narrative of Acadian resistance to the challenges they had faced throughout their history, but particularly over the previous thirty years, starting with the battles for bilingualism and control over their own institutions in the 1960s and continuing with the defence of Acadians removed from their lands during the creation of Kouchibouguac National Park.[91] For Daniel LeBlanc, who assumed a leadership role in the campaign against the causeway in the late 1990s, Acadians came to see the destruction of the Petitcodiac with "a sense of injustice" that needed to be resolved.[92] As Michel DesNeiges explained in 1998:

> We are today at a crucial turning point in the history of both the Acadians and the Petiticodiac. Indeed, their prospects are linked. As societies across the globe define themselves in terms of technological progress, an opening up to the world, and the restoration of the environment, the Acadian people need to keep pace ... What message are we communicating about ourselves to the world if we allow the Petitcodiac to die?[93]

More specifically, DesNeiges observed that "saving the river is the most important issue for the young Acadians of this region."[94] Indeed, as the ranks of those opposed to the causeway grew in the 1990s, young people, particularly Acadian youth, played a key role. For instance, the first act of public resistance to the 1992 decision to keep the gates closed came from two dozen university students from across the province, who obstructed traffic on the causeway in order to distribute leaflets to those driving by, while chanting: "Open your mind, open the gates."[95]

This connection of Acadian youth culture with the anti-causeway movement was reinforced in the song *Petit Codiac,* written in 1993 by Yves Chiasson of the Acadian band Zéro Degré Celsius and carried to a large audience in a recording by the popular singer-songwriter Zachary Richard. The song did little more than name places, most of them in the Acadian sections of New Brunswick, an act of resistance in its own right, claiming space for a people who do not control any territory. The theme of resistance was strengthened in the chorus, which placed Jackie Vautour, the leader of resistance at Kouchibouguac, in the pantheon of those who had opposed authority against all odds: Crazy Horse, Beausoleil (Broussard), and Louis Riel.[96] But the song began, and received its title, from a plaintive cry:

> Petit Codiac,
> Rivière jaune,
> Petit Codiac,
> Rivière jaune,
> Petit Codiac,
> Rivière bleu.

Chiasson explained to me that although the song was not written in the context of the movement against the causeway, it was adopted by the structure's opponents and played in all recorded rallies against the gates. In the process, according to Daniel LeBlanc, the song played a key role in mobilizing young Acadians.[97]

Young people were also brought into the anti-causeway movement through the schools. Toward the end of the 1993–94 school year, several students, writing to the environment minister, explained how they had been made aware of the situation by "two people who, for their summer job, came to our school to speak to us and to show us some slides about the Petitcodiac River. It was very interesting."[98] As a result, petitions with hundreds of students' names were submitted, invariably in French and from either Moncton or Dieppe, its mostly Acadian suburb. Along the

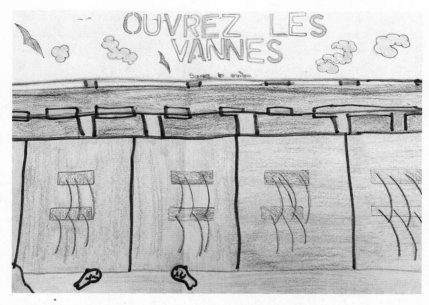

FIGURE 4.6 Student art against the Petitcodiac Causeway, 1997. Artwork was created by a minor, whose name has been withheld. *PANB, RS 748/5270-P/ 0-97-45, 19375.*

way, the minister received various types of appeals. One student, after sending along a letter and a poem, added a postscript: "I am very serious. I am afraid that the river is really going to disappear which would cause the extinction of many animals. I hope you are going to take this seriously. Thank you, and remember! Extinction lasts forever, and forever is a very long time."[99] Another student, in response to a further effort in 1997 to mobilize youth support, sent in a drawing urging the government to "open the gates"[100] (see Figure 4.6).

THE CAMPAIGN TO MOBILIZE the population against the McKenna government's decision to keep the gates closed saw almost immediate success. A draft memo to the provincial cabinet in late 1992 noted how "over the course of the past year the nature of the issue seems to have shifted. Originally, it was phrased as whether to protect the perceived interests of the twenty-one residents who live on the headpond or whether to achieve the environmental benefits that would accompany opening the gates." The wording here was significant, quantifying the limited number of property owners and referring to their "lake" as a "headpond." The memo went on to explain that "the question now is less whether or not the gates *should* be opened, but rather how to control the costs for erosion protection that would be needed if the gates were opened." Following a reassessment of the situation, there was now a consensus in Fredericton "that opening the gates is in the best interests of both the economy and the environment. All of the departments are agreeable to revisiting the issue."[101]

But "revisiting" the situation did not mean that McKenna was now willing to permanently open the gates. In fact, in his notes for the cabinet meeting to reconsider the matter, the premier observed that "the credibility of this government's response to the Round Table Sustainable Development Action Plan will be sorely tested if not willing to at least re-examine the issue." He insisted that he was "not asking for a reversal of the decision to close the gates," which made his government eager to seize on a proposal in 1994 by Alyre Chiasson, a professor at the Université de Moncton, to carry out a test that would provide data that could be useful for a permanent opening – somewhere down the line.[102]

In the short term, however, Chiasson was only proposing a seven-month experiment that would see one gate opened in the spring during low tide, allowing the water to drain from the headpond; then, as the tides came back upstream, the gate would remain open only until the water level in the headpond was 2.5 metres deep, as opposed to the 6 metres that had previously been the case. In the process, enough water would pass upstream to fill the pre-causeway river channel. Cutting off the tide – "clipping" it, to use Chiasson's expression – would limit the force of the water moving upstream, reducing the need to make costly investments to return to the tide protection that had existed before 1968. Chiasson hoped that fish passage would be improved as a result, although DFO remained concerned "that the restricted time period for opening the gates on the incoming tide (estimated at around fifteen minutes) may severely limit the ability of adult returning fish to migrate upstream to spawn."[103]

Chiasson did everything he could to downplay the significance of the test. He noted that "people think there's a permanent decision being made here to open the causeway gates ... That's wrong. What's being conducted is a trial experiment. It's only a commitment to do a trial run with a gate open, not to do away permanently with the lake."[104] Nevertheless, it was difficult to deny that even this half measure would compromise the headpond as it had existed for over a quarter-century. Chiasson recognized that "people along the lake have this feeling of being left empty-handed." Perhaps a bit naïvely, he suggested that "if the lake does get permanently drained after some future decision, we could replace the value of the lake with nature trails, or ponds for ducks. We could replace the lake with something else – another kind of place that would be good to look at."[105]

Now faced with the McKenna government's backtracking on its promise to keep the gates closed, the residents along the lake mobilized to prevent the Chiasson experiment from taking place. They staged a "Save Our Lake Rally" (the first of many) in August 1995, and only a few months later the Lake Petitcodiac Preservation Association (LAPPA), which became the voice for those wanting to keep the gates closed, was formed. At roughly the same time, the *Times & Transcript*, recognizing that the lake residents were digging in their heels, observed: "To conclude that a single group has the right to hold hostage part of the river for its own satisfaction is an indefensible position in the face of the potential benefits to all."[106]

In fact, LAPPA and its supporters did manage to delay Chiasson's experiment, ultimately making it impossible to carry out the test when it was finally scheduled to begin in 1998. By filing a court challenge (which was rejected), they prevented the experiment from starting sufficiently early in the season to yield the desired results.[107] Along the way, many of the same people who had been active in the movement since the 1980s made the same arguments once again – that the headpond, after thirty years, was now part of "nature" and so deserved to survive, especially if one doubted that the fish stocks would return. As one LAPPA member put it:

> I am very concerned about the decision of our Government to have the gates opened at intervals for the clipping of the tide experiment. This will definitely ruin the lake and the surrounding eco-system. I've had the opportunity of seeing the beauty of this artificial lake and its surrounding area with its abundance of birds and wildlife ... It has the potential of all kinds of water activities if it were not for the threat of destroying this lake by opening the gates ... In my opinion, we, the members of LAPPA, are the true naturalists, environmentalists. We are trying to save the lake and the surrounding eco-system, not destroy the lake just to satisfy the wishes of people whose intentions might be good but unrealistic if they think the river will go back to the way it was before the causeway.[108]

But there was also something new in the letters sent to Fredericton to prevent the clipping experiment from taking place. Parallel to the insistence by some Acadians that they had a particular connection with this landscape, there were lake residents who pushed back, rejecting the idea that Acadians had a role to play in the debate. Take, for instance, the letter sent by Mary and Blair Dolan from Riverview to their MLA in 1997:

As far as we are concerned, this has become a French-English debate with the Moncton University profiting from New Brunswick Government grants for tests, surveys, etc. We English residents who live on the Lake want to keep the Lake for recreation, boating, snowmobiling, and beautification of surrounding properties ... We, as tax payers on the Lake, are tired of these people at the University of Moncton and the people of Dieppe running our government.[109]

The Dolans' reference to Dieppe was a gratuitous swipe at the rapidly growing, mostly French-speaking, middle-class suburb of Moncton, which represented Acadians in a light that may have discomfited some English-speakers. As for their unhappiness with the Université de Moncton, where Chiasson was a professor, the institution had been in its own right a hard-won victory for Acadians in the 1960s, providing them with the means of developing their own expert knowledge. In reaction, some tried to delegitimize the Acadians' newfound expertise and to question the value of the research being done at the university. In that spirit, Ronald and Beryl Gaskins, who were constantly writing to Fredericton on the issue of the causeway, viewed the clipping initiative as little more than "a make work project for the University of Moncton."[110]

The ability of LAPPA to force the cancellation of the tide clipping experiment in 1998 only added to the sense of powerlessness among Acadians such as Suzanne Doucet, who wrote in French to McKenna that "because of my Acadian roots, I know the importance that the Petitcodiac watershed had for my ancestors. It pains me to see how a small group of residents unfortunately have the power because of their money to determine that the causeway gates should remain closed so that they can keep their artificial lake." This situation was particularly difficult for her to accept since "all of the studies by researchers have shown the negative effects of keeping the gates closed."[111] While I found letters written in English in the Provincial Archives of New Brunswick that spoke to both sides of the causeway debate, I did not find a single one in French that supported the continued closure of the gates.

WHILE THE CLIPPING experiment raised passions, it ultimately produced meagre results. After being terminated in 1998, it ended abruptly once again the following year when dry conditions made it impossible to refill the headpond, even to the targeted 2.5-metre mark, after it had been drained. As a result, the project was declared over, the gates were closed, and the reservoir was filled once again to its normal 6-metre elevation. On the face of it, nothing had changed, but that would not be a correct reading of the situation, as those who wanted to restore full tidal flow (and not the half measure of the clipping trial) now pushed even more energetically to get the job done.

The new energy in the anti-causeway campaign mirrored the growth of the environmental movement across the globe. Following the first Earth Day in 1970, the movement only became stronger in the face of a succession of environmental disasters, including revelations regarding Love Canal and the partial meltdown of the Three Mile Island nuclear reactor in the late 1970s, as well as the large-scale loss of life in the 1980s caused by a gas leak at a Union Carbide facility at Bhopal, India, and the nuclear accident at Chernobyl in the Ukraine.

In this context, environmentalists successfully pushed for changes, in some cases rolling back projects – such as the damming of the Petitcodiac – that had resulted in the redirection of water. Take, for instance, the case of massive water diversion in the American West for irrigation purposes. Writing in 1986, Marc Reisner could have been describing the Petitcodiac watershed:

> By erecting thirty thousand dams of significant size, [the federal government] dewatered countless rivers, wiped out millions of acres of riparian habitat, shut off many thousands of rivers of salmon habitat, silted over spawning beds, poisoned return flows with agricultural chemicals, set the plague of livestock loose on the arid land – in a nutshell, they made it close to impossible for numerous native species to survive.

But writing only a few years later, in an afterword for a revised edition of his book *Cadillac Desert*, Reisner had a very different perspective:

> It didn't seem possible when I began writing this book, but by now it is beginning to seem plausible after all. After damming the canyons and dewatering the rivers in order to spill wealth on the land, we are going to take some of the water back, and put it where it really belongs ... There is more talk of deconstruction than of construction: of minor dams demolished, of big dams made "environmentally sound," of marginal acreage retired and water returned to its source, of flows bypassing turbines to flush salmon and steelhead out to sea.[112]

As for how such change had occurred, Reisner pointed to the aggressive use of the courts by environmentalists who insisted that governments enforce the laws that were on the books, some of which had been enacted in the first place as a result of lobbying by environmental activists. As he put it, "law has been the ignition," an observation that also pertained to the ultimately successful campaign to permanently open the Petitcodiac Causeway's gates.[113] Crucial in this context was the role played by the Riverkeeper movement, which emerged in the 1960s as part of a legal battle by fishers in New York State. It challenged polluters in the courts for having damaged the Hudson River, which, like the Petitcodiac, was a tidal estuary where fish stocks had been decimated. Robert F. Kennedy Jr., who became Riverkeeper's chief prosecuting attorney in the 1980s, described the organization's approach, writing with John Cronin, with whom he created the organization: "Its brand of aggressive and litigious environmentalism saved the [Hudson] river, and its victories generated a wealth of environmental jurisprudence that communities across America now use to protect their own public resources."[114]

But the Riverkeeper movement's influence did not end at the Canadian border, and in 1995 Kennedy brought his star power – and his expertise – to Moncton, pledging his movement's support for the restoration of another river, in the process providing the anti-causeway campaign with international visibility.[115] His visit was encouraged by Michel DesNeiges and his fellow environmental activists at the Université de Moncton, who had founded Écoversité in 1993. DesNeiges recalled how Kennedy "put on these huge fireman boots ... and wandered off into the muddy banks of the Petitcodiac – and then he sank up to his waist in the mud." A photo from that adventure is shown in Figure 4.7, but what DesNeiges particularly recalled in interviews at the time and in one with me in 2019 was how Kennedy encouraged his group to be more confrontational: "It was a watershed moment. I do believe it was the defining moment in this issue. That crystalized it for us. We had not been a confrontational group until then. That's when we became militant."[116] Feeling the wind in their sails, DesNeiges and his colleagues worked, with support from the Riverkeeper leaders in the United States, to create Canada's first chapter in 1999 (in addition to the more than twenty in the United States), with Kennedy once again on hand to mark the moment.

With the tried-and-tested Riverkeeper techniques in hand, the Petitcodiac chapter quickly moved into action, appointing Daniel LeBlanc as its executive director. LeBlanc came to the job with significant experience in dealing with large and difficult problems, having worked for the

FIGURE 4.7 Robert F. Kennedy Jr. at Moncton, along the Petitcodiac River, with Julia Chadwick and Michel DesNeiges, February 24, 1995. *Courtesy of Michel DesNeiges and Pierre Landry.*

United Nations High Commissioner for Refugees in a variety of contexts; closer to home he was involved in creating the Commission de commémoration internationale du Grand Dérangement, so that the far-reaching dimensions of the Acadian deportation might be recognized.[117] On assuming his new position, LeBlanc turned to tactics that Riverkeeper had successfully employed in the United States. As he put it, his organization's "strategy was to stay focused on the law," an approach that came naturally to Acadians, who – as LeBlanc explained to me – had become accustomed to taking legal action in order to have their rights respected.[118] With legal opinions in hand indicating that the federal government was in violation of its own regulations regarding fish passage, Riverkeeper pushed DFO to act, so that in 2000 it appointed Eugene Niles, a former fisheries official, to review all existing studies, consult with interested parties, and make recommendations.

In the report he submitted a year later, Niles correctly recognized that "no one option will satisfy the wishes or strongly held views of all stakeholders." He particularly stood to antagonize the defenders of the lake when he recommended that the two levels of government should work together toward carrying out "a full environmental assessment based on ... the construction of a partial bridge in the Petitcodiac River Causeway," in the process entirely removing the gate structure and permanently draining the headpond. Since his mandate was to respond to DFO's responsibility to restore fish populations in the river, he opted for the solution that might bring "fish passage [to] as close to pre-causeway conditions as possible."[119]

In a sense, there was nothing surprising about Niles's recommendation, since the science had supported this course of action for decades. Nevertheless, as he correctly anticipated, residents along the headpond reacted badly to the report. Victor McLaughlin, whom we met earlier, decided not to attend a meeting to discuss the findings: "In spite of a Beta-blocker to control an arrhythmia, I think that my cardiologist would think it best that I don't."[120] And while Daniel LeBlanc welcomed the report, he returned to the threat of legal action to compel the pertinent governments to carry out the environmental impact assessment (EIA) that the Niles report called for before any permanent change could take place. As LeBlanc explained to the minister of fisheries and oceans:

> We are now reactivating our organizational focus on the legal challenge with the intention of commencing action against the Government of Canada ...: The absence of any evidence from your Department or the Province to follow through with the recommendations of the Niles review leaves us with the no other recourse than to pursue this course of action.[121]

The EIA finally began in 2002, with results only released to the public three years later, by which time the Petitcodiac had been declared "Canada's Most Endangered River."[122] This designation was bolstered by reporting in the local media that highlighted visuals reflecting the degradation of the river: "Scenes of stranded dolphins beating around in silt in an attempt to reach water that wasn't there. A moose unable to budge as he stood hopelessly buried up to his chin in mud. People with cameras waiting to take pictures of a tidal bore that never reached the causeway."[123] The EIA report reinforced those images, concluding that the continued closure of the gates had been responsible for "many costs associated with the significant negative environmental effects [that were] predicted," and of course, there was also "the ongoing violation of the Fisheries Act." The report made it clear that only the permanent opening or removal of the gates could bring about "the restoration of fish passage and overall ecosystem benefits," such as dislodging the sediment.[124] In essence, this was a death warrant for the headpond.

The EIA explored various options to achieve the desired end, and concluded that the course of action most likely to allow fish passage to return to the Petitcodiac in a cost-effective manner was the removal of the gate structure and construction of a 280-metre bridge in its place. In making this suggestion, the EIA rejected a more modest option that would have resulted in a shorter bridge span, with less tidal flow and correspondingly less erosion of the sediment that had built up over the previous forty years; this silt needed to be washed away for the river to return to something approaching its pre-1968 state. At the same time, the EIA also passed on a more radical option that would have allowed for a much longer bridge, but at twice the $54 million bill for the preferred option and with the danger of creating so much tidal flow that the former Moncton Landfill, immediately downstream from the causeway, might be disturbed.[125]

The EIA made the case for what it saw as a safe choice, what Daniel LeBlanc called "a middle-of-the-road solution," but this was not how the report was perceived by property owners along the headpond.[126] At a meeting to allow community feedback, individuals connected with LAPPA challenged both the competence and the impartiality of the individuals behind the EIA. For instance, Jim Sellars, who owned property on the Moncton side of the headpond, within a stone's throw of the causeway, pointed to "the lack of substantive research, the lack of due care and diligence in considering all the options, the lack of careful accounting, the lack of proper evaluation of the issues ... As somebody said, 2355 pages I believe is what it is in English. If you have to use 2355 pages to tell this story, there's something wrong with the story. And obviously you're spinning it and winding it up in such a way as to pervert the final result."[127]

It was ironic that the report's opponents dwelled on its lack of scientific rigour, considering that 130 studies had been carried out, nearly all calling for the opening of the gates.[128] Nevertheless, another participant (not identified in the transcript) in the public hearing over the EIA's conclusions observed:

> I've been appalled by the ignorance of people from the University of Moncton and from University of [New Brunswick] Environmental Studies department in Fredericton. When it comes to how the ecosystem works, I say we are the only people [who know] ... We shall not be intimidated when ... people come along with big scientific degrees and try to intimidate us and bully us into solutions they don't pay for, but in the end the working stiff has to pay for.[129]

As we have seen at other moments during the controversy, LAPPA supporters turned the scientists' expert knowledge into a liability, preferring instead their unsubstantiated claims that they had a superior understanding of their third-nature landscape.

With the release of the EIA report, the ball was in the court of the New Brunswick government, led after the 2006 provincial election by Shawn Graham's Liberals, who heard an earful from those still clinging to the idea that the headpond could be saved. Having seen previous governments drag their feet in the face of such lobbying, Petitcodiac Riverkeeper went to court in July 2007, insisting that DFO "force the owner of the causeway, the Province of New Brunswick, to allow for the unimpeded passage of fish in accordance with the Fisheries Act." As Michel Desjardins, the chair of the organization's board, explained, Riverkeeper had been ready to file a month earlier, "but there were rumours that the Province had chosen one of the options and was in negotiation with the federal government, so we decided to give them a little leeway. Well today the hourglass is empty and their time is up."[130]

It is impossible to know whether this further threat of legal action made the difference, but a month after Riverkeeper's court filing, Fredericton announced that it had accepted the EIA's preferred option, and committed to the construction of a bridge to replace part of the causeway. Graham's decision drew a wide array of responses, as would be expected given how the issue had polarized the region for decades. On the one hand, there was the resident along the headpond who told the premier how his "daughter cried when she saw the newspaper this morning [reporting the decision]. She asked where will the geese swim. We will miss the wildlife, the boating, the snowmobiling as well as the view. The water will now be polluted with city sewer water. We will be left to look at a muddy, smelly fly trap."[131] At the other end of the spectrum, there was the apartment-dweller from Riverview (so not someone with waterfront property) who thanked "Premier Graham for his perseverance and vision in bringing about the permanent opening of the controversial gates on the Petitcodiac River. At last we have a politician with balls. I remember years ago when Premier Frank McKenna was on the verge of doing the same thing but backed down due to extreme local political pressure."[132]

With $20 million of provincial funding, the first phase of the restoration project, leading to the opening of the gates, began in 2008. Crucial to this phase was the construction of dykes and aboiteaux upstream from the causeway, so that land would not be flooded when tidal flow returned. Such structures were now needed precisely because the MMRA had not rehabilitated the ones already in place, which were made redundant by the causeway; and those pre-existing structures were frequently removed after 1968 by farmers such as Jim Reicker, who in the process was able to secure additional cultivable land on his property thirteen kilometres upstream from the causeway.[133]

During the restoration process, remains of earlier, sometimes Acadian, dykes were unearthed, a reminder that what was being restored was second nature, a human construction every bit as much as the causeway being removed. To be sure, there was always something jarring about the claims of residents along the headpond that they were defenders of nature, at the same time that the causeway was decimating fish stocks. However, there was also something discordant about two of the leading anti-causeway activists, Gary Griffin and Julia Chadwick, writing about how the Acadian construction of dykes had had "little effect on the creatures which depended on the uniqueness of the Petitcodiac for their existence. The dyke structures were constructed in such a way as to allow fish and wildlife passage."[134] Writing in this manner, Griffin and Chadwick naturalized the dyked landscape, ignoring the first nature that had preceded it and was destroyed by it. In the process, they obscured how the salt marshes supported species that had now lost their source of nutrients, a process that paralleled the compromise of fish stocks by the causeway.

In this context, in his 2001 report regarding the options for the Petitcodiac, Eugene Niles made the astute observation that the removal of the causeway would result in returning the landscape to some earlier form of nature. Recognizing that there had been more than one form, he wondered "how far back one should attempt to turn the clock in restoration efforts. For example, should we attempt to restore all the former wetlands and salt marshes that have been drained and dyked over the past three or four centuries and are now being used as productive farmland?"[135] In the end, however, no one seriously proposed returning to first nature, and so work proceeded to prepare the river for return of tidal flow, with the new dykes and aboiteaux in place, in the spring of 2010.

Along the way, farmers lost the very land they had gained with construction of the causeway, as dykes and aboiteaux were needed once again. Wiebe Leenstra, who owns a large dairy farm just upstream from Jim Reicker, was philosophical about the loss of eight acres, which he figured was balanced out by other land that would now be better drained and more productive.[136] Reicker was less sanguine, however, making the case to Fredericton in 2009 about the land he had lost: "My wife and I have lived on the Salisbury Road, for a few months short of forty years," meaning that they arrived just after the closure of the river. Referring to new legislation that gave the province the power to enter his land to construct the dykes and aboiteaux now required due to the imminent opening of the causeway gates, he continued:

> The government intends to enter our property, use approximately 2000 feet by 100 feet to construct a dyke and destroy the river as we have always known it. Because of the dyke, we will have no further use of the river for recreation – we won't even be able to see the water. These dykes will landlock one of our properties. They tell us that this factor will have no effect on the use of our property or the re-sale value and that we will not be compensated in any way. This does not seem reasonable to us.[137]

Reicker was not alone in resenting the end of the headpond, and LAPPA predictably tried one last-ditch effort to have the courts block the opening of the gates, but this was no more successful than previous attempts. In dismissing the demand for an injunction, Justice Paul Creaghan of the New Brunswick Court of Queen's Bench observed: "I think LAPPA has also concluded that delay is the only avenue that might result in a change of government policy that could maintain the headpond they have enjoyed for some forty years ... The Plaintiff does not advance a legal basis to stop the project."[138]

With the last obstacle cleared, the gates were opened permanently in April 2010, with more than 300 people, on both sides of the issue, on hand. One newspaper reported that "there were parts of the crowd standing to raise their arms in a rendition of 'the wave' as the gates opened to cheers. Other parts of the crowd booed in attempts to quell the elation."[139] The long-time defenders of the "lake" then returned home to see the draining of the reservoir, where large stretches of mud now led to the river channel, the boat ramp from what once was the marina having become a road to nowhere (see Figure 4.8).

By contrast, for those who had worked to open the gates, there was considerable excitement over the transformation of the river, which began almost immediately. Within a month, the Petitcodiac Fish Recovery Coalition installed a fish trap upstream from the gates, near Salisbury, now again the head of tide, where in one day they caught and released 5,000 gaspereaux. As former Riverkeeper Daniel LeBlanc put it: "I hope people understand that this is history they're witnessing."[140] Results over the years that followed provided further cause for optimism. In 2015, Edmund Redfield, who was overseeing the coalition's fish trap, explained:

FIGURE 4.8 Ramp leading to the former marina on Lake Petitcodiac, Riverview, NB.

Many native species populations that we hoped would increase are indeed doing so, some far more than expected, while the numbers of non-native invasive species are declining ... The results of the latest monitoring suggest that, after decades of absence or reduced numbers, increases in quantities of striped bass, American eel, rainbow smelt and Atlantic tomcod are evidence that many native fish populations in the Petitcodiac River are rebounding.[141]

The return of fish stocks was not the only change to the river. There was also evidence that tidal flow was washing sediment away, allowing the river to widen. The Petitcodiac Fish Recovery Coalition observed in 2012, only two years after the gates were opened, that the river had already widened by 20 metres one kilometre downstream from the causeway. This was a long way from the one-kilometre width before the causeway, but it was an increase of 25 percent over the 80-metre width at the time the gates were opened. There were even more dramatic results further downstream, where – at ten kilometres from the causeway – the channel had widened by 350 metres.[142]

With the expanded river, there was also the return of the tidal bore, announced by the surfers who now looked to the Petitcodiac for the perfect wave. The Americans who came to Moncton to break a world record in 2013 were joined by Antony Colas from France, who had surfed tidal bores around the world. Colas knew that surfers had been drawn to the Petitcodiac prior to the construction of the causeway, and was waiting for just the right moment to surf the river now that the gates had been opened. He knew that the moment had arrived in June 2013 "when I came across a video clip announcing the coming of the Supermoon," what Yassine Ouhilal described as a full moon, which was "50,000 km closer to Earth than normal in elliptical orbit. This proximity resulted in a visibly 'bigger' moon, a third brighter, creating tidal swings 20% greater." Colas knew that "this would be an amazing surfing opportunity. My heart skipped a beat."[143]

Inspired by the supermoon, Colas and Americans Wessels and Whitbread independently made plans to arrive in Moncton in late July to be on hand for the next full moon, and they were joined by still others. Colas described July 24 as "an unforgettable day," during which there were at one time eleven surfers in the water, "being watched by an enthusiastic crowd, including some who were old enough to have known the bore before 1968. They had tears in their eyes, to see the bore again become a public spectacle."[144]

During his time in Moncton, Colas was assisted by Daniel LeBlanc, whose perseverance regarding the opening of the gates had made surfing on the river possible. LeBlanc ultimately became involved in surfing himself, spurred on by the excitement in July 2013. Since then, he has surfed the river over sixty times and has been joined by surfers from all over the world who come to Moncton for an adventure. All of this was significant for LeBlanc as it helped to "change the public's perception of the river," which little more than a decade earlier had been declared all but dead.[145]

And for all the surfers, there are even better days ahead. While the Americans had their record-breaking ride halted by the gate structure, surfers can now continue further upstream, thanks to the construction of the bridge to replace the obstruction, in the process allowing even greater tidal flow. Following further foot-dragging after the opening of the gates, the presence of Liberal governments in both Fredericton and Ottawa led to a commitment in 2016 of nearly $62 million, divided fairly evenly between the two governments, to construct the partial bridge.[146] The completion of this project in late 2021 brought to a close a story that dragged out over more than forty years, in the end undoing one of the MMRA's legacies and allowing the return of second nature.

FIGURE 4.9 Petitcodiac salt marsh downstream from Moncton, 2003. *Courtesy of Michel Rathwell.*

THE RETURN OF THE MARSH

Throughout most of its history, the MMRA repaired or constructed dykes and aboiteaux to ensure that salt marshes would be cut off from the tides of the Bay of Fundy, in the process draining them and allowing arable land to take their place. But when the agency, near the end of its mandate, installed large dams across the Petitcodiac and Avon Rivers, the legacy of its actions led to the return, however unintended, of the marsh. In the case of the Petitcodiac, as large quantities of sediment built up, narrowing the channel downstream from the causeway, conditions were ripe for salt grasses to reappear, in an ironic twist recreating something of the landscape that had existed prior to the Acadians (see Figure 4.9). As anti-causeway activist Gary Griffin put it in the early 1990s: "By 1991 we have lost 87% of the estuary which was previously swept clean twice a day by the tides. All this tidal area has been replaced with, guess what, hundreds of acres of salt tidal marsh ... on both sides of the river stretching twenty-five miles downstream."[147] Griffin and his colleagues had nothing positive to say about these marshes, which they viewed as forming part of the degradation of the Petitcodiac, and they hoped that with the construction of a bridge and the free flow of the river, the sediment would erode over time, and with it the marsh grasses.

FIGURES 4.10–4.12 Accumulation of silt downstream from Avon Causeway: 1966, 1981, 2003. Immediately upstream in the post-1970 photos is Lake Pesaquid. *Aerial photography from the Nova Scotia Aerial Photography Database. Crown copyright 1966, 1981, and 2003, Province of Nova Scotia. Used by permission of the Department of Service Nova Scotia, and Internal Services License #VA6211A. All rights reserved.*

By contrast, a very different dynamic developed with regard to the similar buildup of sediment, and subsequently marsh grasses, immediately downstream from the Avon Causeway, as indicated in Figures 4.10 to 4.12. Danika van Proosdij, who has written extensively on the evolution of the Avon mud flat, described how "sediment began accumulating rapidly in the vicinity of an existing mud/sand bar [that] appears to have been present since 1858." Aerial photographs show the growth of the mud flat after the installation of the causeway in 1970, followed by the appearance of "salt marsh vegetation ... on the exposed mudflat surface around 1981 ... After 1992 the rate of colonization by *Spartina alterniflora* increased exponentially as the vegetation became firmly established on the mudflat

(4.11)

(4.12)

surface ... By the summer of 2005 almost the entire suitable mudflat surface had been colonized."[148] The Windsor Salt Marsh (so named because of the nearby town) continues to grow by nearly five acres per year, with some of its grasses measuring 121 centimetres (nearly four feet) in height.[149]

While salt marshes grew as a result of causeway construction in both the Petitcodiac and Avon estuaries, in the latter case they have come to be seen not as a nuisance but rather as a very positive, if unplanned, by-product of the MMRA's initiative. Writing with Sarah Townsend, van Proosdij explained how the Windsor marsh drew attention to the "positive impact of the causeway, the establishment of a thriving downstream salt marsh ecosystem," which has become an important feeding ground for shorebirds.[150] In this context, the marsh became a public good that needed to be defended in the face of the calls by some that the causeway should be replaced by a bridge, echoing the demands with regard to the Petitcodiac. While full tidal flow was the goal in Moncton, at Windsor, as one study explained, "the removal of this causeway would probably eliminate what appears to be one of the most productive marshes in the Bay of Fundy system."[151]

In this context, the 2019 Nova Scotia Tourist Guide referred to the Windsor Salt Marsh as a "top photo opportunity," providing a "180 degree view of the salt marsh, the wide red mud flats, and the many migrating shorebirds that feed here during their journeys" (see Figure 4.13). As Lisa Szabo-Jones has observed, "the town celebrates this wonder with a dedicated website and a marsh webcam." She cautions us to remember, however, that "Windsor Marsh is not proof that nature can bounce back after major human disturbance. In this instance, all the conditions were in place for a marsh to grow," conditions that included the presence of raw sewage, which in a further ironic twist helped stimulate marsh growth because it contained high nitrate concentrations.[152]

IN ORDER FOR THE Windsor Salt Marsh to be perceived so positively, local residents needed to overlook other legacies of the causeway, most notably the obstacles that it created for the movement of fish up and down the Avon River. While the Petitcodiac River had been unarguably a fishers' paradise prior to its obstruction, this situation was much more ambiguous with regard to the Avon. As we saw in the previous chapter, the federal government did not require the inclusion of a fishway, largely because of reports that dams installed upstream decades earlier had already destroyed much of the river's fish stocks. Nevertheless, there were individual accounts

FIGURE 4.13 Windsor Salt Marsh. *Courtesy of Lisa Szabo-Jones.*

from both local residents and fisheries officials suggesting that the situation was not so dire, and this suggestion was supported by various incidents in the years following the closure of the causeway.

In one such case, when the causeway's gates were opened in 1986 to permit repairs, the local newspaper reported that "a very interesting thing happened. Thousands of gaspereaux made their way back up the water system. Alert fishermen had a field day. One person dropped into the *Hants Journal* office and said people were pulling fish out with buckets." The newspaper continued: "Why is this significant? For years, since the causeway was built, West Hants Wildlife Federation officials have advocated the installation of a fish gate in the causeway. This would allow all of the migratory fish a passage into the rivers and streams."[153]

Following a request from the Wildlife Federation, two DFO scientists investigated "the possibility of providing suitable fish passage at the Avon River Causeway." Vern Conrad and Rick Semple examined the situation and concluded that most species could flourish if the gates were carefully manipulated. In the case of the gaspereaux observed upstream from the causeway in 1986, the scientists explained how the gates were opened each spring to permit maintenance. When the opening "coincided with the upstream migration of alewives [gaspereaux], large quantities of fish were present in Pesaquid Lake [the headpond] and the tributary streams." The authors theorized that with proper manipulation of the gates, alewives could then "spawn and rear" in the headpond, and they were similarly confident that "if suitable fish passage conditions could be provided at the causeway an increase in sea-run trout could be expected in the West Branch [of the Avon River] and in the tributary streams of Pesaquid Lake." In addition, Conrad and Semple proposed stocking an upstream branch of the Avon with fingerlings (juvenile salmon) each fall and regulating outflow from the gates during low tides in summer to allow them to escape.[154]

If the Wildlife Federation had hoped that the DFO study would lead to the construction of a dedicated fish passage, Conrad and Semple disappointed them, as they were careful to make recommendations that limited fish passage to avoid salt infiltration into Lake Pesaquid, insisting for instance on opening the gates only when the tide was out. Such half measures were ultimately rejected with regard to Lake Petitcodiac, which came to be viewed as disposable, linked as it was to high-end properties. By contrast, Lake Pesaquid played a very different role, situated adjacent to the downtown area of Windsor, which was now protected from flooding. The headpond serves as a recreational area for the town, symbolized by the Pisiquid Canoe Club, created shortly after the river was closed, and the town's annual Pumpkin Regatta, established in 1999, which supports the Children's Wish Foundation.[155] Each year, pumpkins, some weighing over 500 pounds, are hollowed out and brightly decorated before being paddled across the lake, with townspeople on hand in significant numbers to take part in an event that, in a sense, celebrates the headpond.

The dynamics of this situation changed, however, as plans began emerging in the early 2000s for the province to expand the highway that ran over the tidal gates, which would have to be rebuilt in the process, in part because the MMRA's structure was coming to the end of its life; and while DFO had allowed the construction of the original causeway without any facilities for fish passage, it was unlikely to do so again.[156] Such facilities, as we saw in the Petitcodiac context, could take various forms, ranging from the inclusion of a dedicated fish passageway (while still maintaining the integrity of the freshwater lake) to the replacement of the gate structure with a bridge to permit full tidal flow, as was done in Moncton.

In that context, the group Friends of the Avon River emerged to secure the same solution that had been achieved by Gary Griffin, Daniel LeBlanc, and their colleagues in the Moncton area. Sonja Wood, a leader of the group, described in 2003 "how 20–40,000 gaspereaux were caught at the causeway gates in 2002 and died as a result." As for the Windsor marsh, she did not see a productive landscape but rather echoed the anti-causeway activists in Moncton in seeing only "a disgusting desert of mud."[157] Viewing the causeway as "an environmental disaster," she recognized that "the damage has already been done, now we have to repair it or we will see the same thing that occurred on the Petitcodiac River in New Brunswick."[158]

In the years that followed, as the highway project moved toward implementation, Wood continued to push for what she called in 2017 the "Petitcodiac Model." By then, the gates had been opened and the bridge was slated for construction in Moncton, but she was still speaking for only a small segment of the population in the vicinity of Windsor: "I'm not asking to open the entire river at this point ... We're asking for the same help they got for the Petitcodiac; the only thing that works is free tidal flow."[159] But as the province pushed ahead with its highway plans, which included a new structure to control the tides, and put various scenarios on the table for public consultation in 2018, the Petitcodiac model was not even in the mix.

Wood was amenable to an option that called for "controlled exchange of tidal water," which would "provide unimpeded twenty-four-hour, year-round opportunities for the fish to enter and leave their critical habitats." However, most in the community responded to public consultations by supporting a more restrictive formula that would include a fishway but also require "frequent gate closures, intended to maintain status quo lake conditions."[160] The consultations indicated widespread enthusiasm for a design that would "save the lake for business, community, tourism and recreational purposes," a sentiment reflected in signs displayed on many Windsor storefronts when I was there in 2019 (see Figure 4.14).[161]

In this context, the province pressed ahead with its "hybrid" formula, incorporating both tidal gates and fishways. This design did not satisfy Wood and her colleagues, who continued to push for "the uninhibited return of free-tidal-flow." As for the proponents of the plan, they viewed it as "a balanced solution" that would "maintain the lake while maximizing opportunity for fish passage."[162] And while the point was never explicitly raised in these consultations, the plan that is moving toward implementation as I write these lines would also allow the Windsor marsh to continue growing by restricting tidal flow, in the process allowing the further buildup of sediment. Indeed, two of the MMRA legacies, however unintended, Lake Pesaquid and the Windsor Salt Marsh, will continue to exist long after the physical structure that the agency helped build will have been replaced.

FIGURE 4.14 Save Lake Pesaquid campaign. *Courtesy of Robert Buranello, October 2019.*

ONE PERSON'S TRASH IS ANOTHER PERSON'S TREASURE

In spite of the scientific rigour that was at the heart of the MMRA's mission, its operations often had consequences that it did not anticipate. This failure reflected a classic high-modernist preoccupation with the job at hand, to the exclusion of everything else. For instance, in the case of drained marsh near the New Brunswick–Nova Scotia border, the MMRA carried out various calculations to determine the appropriateness of investing funds to reconstruct the dykes and aboiteaux, but failed to account for the absence of sufficient demand for the newly protected, arable land. As for the structures across the Petitcodiac and Avon Rivers, the MMRA's engineers narrowly defined their jobs in terms of blocking the tides, paying little attention to (or choosing to ignore) the consequences in terms of fish passage or the dynamics of disrupting tidal estuaries, in spite (or perhaps because) of the agency's self-conscious involvement with the larger international community of engineers who specialized in working in tidal environments.

Even before the MMRA closed shop in 1970, its projects had already drawn criticism from individuals with expert knowledge of their own, most notably fisheries officers who worked for the same government as the agency's engineers. These same scientists continued to produce studies in subsequent decades, and were joined by a wide array of experts who brought their ecological perspective to bear. But while there was evidence that both the Petitcodiac and Avon Causeways were responsible for the spectacular transformation of the landscape, those most directly affected by the changes, the inhabitants of the two watersheds, responded in very different ways. In both cases, fish stocks were compromised, millions of tons of sediment accumulated, marsh grasses took root, and headponds were created. Third nature came to replace the second nature that the MMRA had originally been tasked with preserving, and yet what was generally viewed as a disaster in one context was largely celebrated in the other.

To be sure, much of the difference can be explained in terms of the extent of the damage that was caused in one case as opposed to the other. There was a reason why the Petitcodiac was labelled the "most endangered river" in Canada, a distinction never suggested for the Avon. Nevertheless, community perceptions of the third-nature landscape also played a role in the process. In the case of the Petitcodiac, a crucial moment in the controversy occurred in the 1990s when Daniel LeBlanc came on board to lead the Petitcodiac Riverkeeper. As Gary Griffin explained to me, LeBlanc insisted that the "lake" should always be referred to as a "headpond," to represent it as unambiguously as possible; and this stark representation was encouraged by the presence of landowners, who could easily be demonized as well-to-do individuals who cared only about themselves.[163] This depiction stuck, particularly within the Acadian community, given that the Lake Petitcodiac Preservation Association was effectively an organization representing English-speaking proprietors, who could be viewed as destroying a river with which Acadians closely identified.

In the end, the social context was crucial to the success ultimately achieved by the opponents of the Petitcodiac Causeway, but it could have ended very differently. As Daniel LeBlanc pointed out, if the structure had been built downstream from Moncton, local residents would have come to view the headpond as an asset, a lake where they could enjoy themselves, much as has been the case in Windsor.[164] In fact, however, circumstances made it possible for Lake Petitcodiac to be represented by residents bass fishing from their motor boats, while Lake Pesaquid has become associated with images of paddlers maneuvering their pumpkin shells (see Figures 4.15 and 4.16).

FIGURE 4.15 Bass fishing tournament on Lake Petitcodiac, pre-2010. *Courtesy of Ann Rogers.*

FIGURE 4.16 Lake Pesaquid Pumpkin Regatta, 2012. *Courtesy of Grant Lohnes.*

Legacies

ACROSS THIS JOURNEY through the Maritime marshlands, we have seen the impact of various human interventions, driven by different types of knowledge, all of which were perceived in relation to the value they might provide, depending on how "value" was defined in the community. There was nothing "natural" about the destruction of the original salt marsh, but it was deemed desirable by Europeans who had a certain idea of what an arable field should look like and what it could provide. Once in place, the drained marsh landscape became naturalized, which for some included the possibility of allowing the tides to occasionally wash over their fields to fertilize them. While the practice of tiding was in sync with the values of farmers who opposed construction of the Tantramar Dam, it was at odds with the mission of the MMRA's engineers, who wanted to control the tides as much as possible, seeing little value in allowing fields to be flooded. And most recently, as both the New Brunswick and Nova Scotia governments close the book in a sense on the legacy of the MMRA, we see very different approaches toward the Petitcodiac and Avon watersheds, reflecting community preferences in terms of the landscape that was valued.

EPILOGUE

Meet the Grand Pre Marsh Body

In April 2019, I attended the annual meeting of the Grand Pre Marsh Body. In an unheated community hall, where boxes of Tim Hortons coffee helped everyone keep warm, roughly twenty proprietors discussed a wide array of issues, such as the upkeep of roads that enabled farmers to travel from their upland buildings to their dykeland, the maintenance of drainage ditches that led water away from their fields, and the setting of assessments the marsh body charged members so it could carry out its work. Discussion also turned to the role of the province in maintaining the dykes and aboiteaux, for which it had been responsible since the closing down of the Maritime Marshland Rehabilitation Administration (MMRA) in 1970. At the end of the meeting, there was a call for nominations for members of the Executive Committee, so that the marsh body could continue to operate as it had since the immediate aftermath of the Acadian deportation.

On the face of it, there was nothing exceptional about a group of farmers meeting to discuss their common concerns, but, as we have seen, the creation of the MMRA hastened the end of the centuries-old practice of farmers managing the dykes, aboiteaux, and drainage channels that enabled them to cultivate land that had once been marsh. In that context, I had considerable difficulty finding the minute books of marsh bodies that had not met for decades; and it was even more difficult to find marsh bodies that were still meeting. This meeting at Grand-Pré reminded me of what it must have been like when more than a hundred such organizations were getting together regularly to watch over their second-nature landscape, putting their local knowledge to work.[1]

Much like the disappearance of rural schools, the passing of the marsh bodies was a further sign that decisions for such communities were now being made in distant places. Robert Palmeter, a farmer whose family had been involved with marshland farming at Grand-Pré since the 1760s (and whose daughter is the current secretary of the marsh body), stressed the continued importance of such an institution in the early twenty-first century: "We still need our local group to provide a voice to the government, as far as what we see needing being done here."[2] And he reinforced this point at the meeting, where he raised a question about the work that had been done by the province over the previous year in terms of keeping up with needed dyke repairs.

As it turned out, no one was in the mood on the evening of the meeting I attended to be overly critical about the province since, as luck would have it, earlier that same day an announcement had been made that the Nova Scotia and federal governments were chipping in nearly $57 million each, in part to raise the height of dykes, most of which had been originally constructed or rebuilt by the MMRA, but which are now challenged by rising water levels due to the combined impact of climate change and the sinking of the region's land mass. One prediction anticipates that the upper Bay of Fundy regions will have high tides nearly a metre higher in 2055, and 1.4 metres higher by 2100, than they were in 2012.[3] The remainder of the funding was targeted for reconstruction of the tidal gates and fishways for the expanded Avon Causeway.

The meeting of the Grand Pre Marsh Body spoke to the best-case scenario for the reconstruction work of the MMRA. Here, and across large swaths of the region, the dykes and aboiteaux were rebuilt, enabling farmers to work fields that remain highly productive to this day. Indeed, the ongoing partnership between the Grand-Pré farmers and the provincial government is an early twenty-first-century version of the sort of mētis that James Scott viewed as a positive force tempering high-modernist excess. Such collaboration was evident across much of the region at the start of the MMRA's existence, as the Amherst-based engineers worked alongside those with local knowledge to reconstruct the dykes and aboiteaux.

In particular, the MMRA borrowed from the new techniques that had been introduced at Grand-Pré, as the marsh body was at the forefront of making the transition from manual labour to machinery. R.H. Palmeter, Robert's grandfather, devised a new method of aboiteau construction before going to work for the MMRA as its initial chief aboiteaux superintendent, bringing his local knowledge with him; Robert's great uncle, Fred Palmeter, also a leading figure in the Grand Pre Marsh Body, provided Léonard Forest with inspiration for Placide, the central character in the film *Les aboiteaux*, which in its own right highlighted the connection between the elderly Acadian's local knowledge and the new scientifically inspired processes being introduced by the MMRA.

In the end, however, Placide was a transitional figure, much like the real-life Ernie Partridge, who brought his local knowledge to the reconstruction of the dykes and aboiteaux in southeastern New Brunswick before he was let go in the early 1960s, by which time the MMRA had largely finished this part of its work. By then, many of the marsh bodies were meeting less frequently, if at all, having concluded that there was little for them to do, particularly as the MMRA intruded on the smallest details of dykeland management.

For its part, the agency, which from the start had questioned the practices of most of the dykeland farmers, viewed the proprietors' apparent lack of interest as proof of the need for a centralized authority to take control. It was in this context that the MMRA shifted its attention from the reconstruction of dykes and aboiteaux to the construction of tidal dams, in the process rendering the marsh bodies upstream irrelevant when the tides were blocked. To be sure, the MMRA was motivated to construct the tidal dams in order to eliminate the costs for repair of dykes and aboiteaux upstream, but cutting out the marsh bodies also figured in its calculations. Take, for instance, the case of the Falmouth Great Dyke Marsh Body, whose lands are located upstream from the Avon Causeway, completed in 1970. While its minute books indicate regular meetings since early in the twentieth century, the farmers did not meet at all between 1969 and 1979.[4]

The Grand Pre Marsh Body stands as the exception that proves the rule, still meeting and suggesting a place for local knowledge long after the introduction of MMRA practices grounded in a high-modernist confidence in the role of experts. Indeed, the survival of the marsh body speaks to its acceptance of modern forms of dykeland management that predated the MMRA and that have continued to this day. In this regard, there is something slightly discordant about the Grand-Pré dykeland's forming part of a UNESCO World Heritage Site celebrating the landscape of Grand-Pré, a site that also includes the most significant Acadian site of memory connected with the *grand dérangement*.[5] As Claire Campbell has observed, in an important study of the use and representation of the Grand-Pré landscape, the UNESCO designation speaks to the drained farmland as a "rural idyll," drawing a "direct lineage between the pré's creation and current [farming] practice, between Acadian settlers and 'their modern successors.'"

Meet the Grand Pre Marsh Body

FIGURE EP.1 Signage posted by Grand Pre Marsh Body.

As Campbell explains, "the UNESCO designation permits and even encourages us to vault over the intervening years of industrial agriculture."[6]

But this public representation was definitely not the self-image presented at the marsh meeting I attended, where tourists' intrusions, walking and biking on the marsh body's roads or on top of dykes, were viewed as obstacles for farmers working on their land. Several years earlier, this concern about outsiders raised some hackles in the larger community, when the marsh body posted signs such as the one shown in Figure EP.1. As Beverly Palmeter put it: "It is to tell you that there is large machinery, equipment being used out on both the dykewalls, as well as the dykeland roads, which are privately-owned."[7] This signage was entirely appropriate given that the proprietors were engaged in operating significant businesses. Indeed, the Grand Pre Marsh Body survived the inroads by the MMRA precisely because it stood apart from those local organizations that disappeared, having failed to make the leap to what Campbell describes as "a global industrial economy."[8]

While Campbell has stressed the transition at Grand-Pré from pre-industrial Acadian farming to the large, highly capitalized units that exist today, she has insisted that we recognize that even the Acadian settlers were involved in an exercise "fuelled by a confidence in the human mastery of nature."[9] Indeed, it was the Acadians who had originally drained the salt marshes, in the process creating second nature, but none of that would be obvious from the UNESCO designation, which naturalized the drained landscape, focusing on how the Acadians introduced dykes and aboiteaux but with barely a reference to the salt marsh environment that preceded it and that had been productive in its own right for millennia.

In the document supporting the Grand-Pré nomination, the only reference to the landscape that the Acadians encountered on their arrival concerned their relationship with the Mi'kmaq, asserting that "when the Acadians began transforming the marsh at Grand Pré, the Mi'kmaq did not prevent them from altering and ultimately removing a vast wetland from the regional resource base." The document asserts that this implicit consent suggested "harmonious relations," although research by William Wicken paints a more complicated picture: "Acadian farming practices forced the Mi'kmaq to redefine how land could be used and what rights accrued to individuals who occupied it. Unlike the Mi'kmaq, the Acadians drastically altered the landscape, building dykes and destroying marshlands that had long been habitats for waterfowl and other animal life."[10]

Nevertheless, the emphasis in the document submitted to UNESCO stressed continuity, so much so that the richness of the "vast wetland" that preceded Acadian settlement was never mentioned by the promoters of the UNESCO submission, a group that included members of the Grand Pre Marsh Body. In that context, there is a direct line between how the Acadians quickly came to see the drained landscape as the norm and how Robert Palmeter described the same environment to me in 2018:

> I think it's been here so long now that yes, it is considered part of nature. I mean, there's a lot of wildlife out here too that do live in our ditches. So there's lots of pheasants and ducks. Geese come in in the fall to feed off of what's left of some of our crops. Yeah it's – today it's just a very natural part of our landscape.[11]

Palmeter was hardly alone, however, in viewing the landscape that he occupied as somehow "natural," scarcely giving a nod to previous versions. As I made my way around the region, traipsing through mud and recording interviews for the documentary film *Unnatural Landscapes*, I heard similar perspectives. For instance, Ernie Partridge, thinking back on his time with the MMRA, observed: "Nature to me, nature is a piece of marsh that has been reclaimed and that ... is utilised to its fullest and is well drained." Similarly, Ann Rogers, the daughter of Dr. Victor McLaughlin, one of the leading opponents of the opening of the gates on the Petitcodiac River Causeway, explained with regard to Lake Petitcodiac, an essential component of the third-nature environment created by the MMRA: "When the lake was there, we couldn't wait till 5 o'clock, when our jobs were done, and we'd come home and run right down to the lake, get on our sailboat, and sail for hours. Or motor for hours. It was just – it was pleasant, it was somewhere to go. Back then, it was a place for nature."[12]

For their part, the MMRA engineers shared the conviction that was evident in the Grand-Pré UNESCO submission that little of significance had preceded the draining of the salt marshes. Speaking before a Senate committee in 1961, just before leaving the agency, J.S. Parker matter-of-factly described the "background detail" to the MMRA's creation, explaining how "over long periods, deposition of silt from [the Bay of Fundy] carried in suspension by tide waters has resulted in buildup of silt beds."[13] Parker made no reference to the valuable grasses produced on those silt beds, and viewed his agency as now responsible for "protect[ing] the areas from flooding during periods of high tide," using whatever techniques were appropriate, either reconstructing the dykes and aboiteaux or building tidal dams. The MMRA implicitly assumed that the salt marsh environment had been unproductive, a point of view that occasionally brought Parker and his colleagues into conflict with farmers who held to the practice of tiding or the harvesting of marsh grasses.

Along the way, as the MMRA deployed its expert knowledge, it was sometimes sensitive to the differing conditions across the region, listening to those who had been involved with dykeland management in the past, individuals such as Ernie Partridge or groups of farmers like the ones at Grand-Pré. More frequently, however, the agency assumed that the farmers had been at least partly responsible for the problems that it found, and treated them accordingly, encouraging the general demise of the marsh bodies. This disregard for local knowledge was most consequential, however, when the MMRA's certainty that it could master the tides resulted in high-modernist disaster, most notably when it contributed to the degradation of the Petitcodiac River. As James Scott explained, when large-scale projects "ignore or suppress mētis and local variation, they all but guarantee their own practical failure."[14] In the end, the deployment of expert knowledge does not have to end badly, but it can, and may well, when the experts are left unchecked in their efforts to reshape the environment.

Notes

FOREWORD: POETS AMONG THE ENGINEERS

1. Ronald Rudin and Bernar Hébert, *Unnatural Landscapes*, available at: http://www.unnatural landscapes.ca.
2. For the exercise, for the attraction of the landscapes, and, no doubt, influenced by the exhortations of A.H. Clark, "Field Research in Historical Geography," *Professional Geographer* old series 4 (1946): 13-22 and Carl O. Sauer, "The Education of a Geographer," *Annals of the Association of American Geographers* 46, 3 (1956): 289–99 declaring that on field excursions, "locomotion should be slow, the slower the better, and be often interrupted by leisurely halts to sit on vantage points and stop at question marks" (296).
3. Andrew H. Clark, *Acadia: The Geography of Early Nova Scotia to 1760* (Madison, WI: University of Wisconsin Press, 1968), remains a definitive statement of its type on the topic and treats the long history of Acadia in detail; quote from page 109.
4. Clark (*Acadia*, 236) suggests a total of 12,600 acres (5,100 hectares) of Acadian dykeland around the Bay of Fundy in 1755, but later (238) observes that the number may be double this (20,000-25,000 acres, or 8,000 to 10,000 hectares), although he says, "the writer feels that figure to be rather high." Two hundred years after the deportation, reports of Acadian reclamation placed the figure as high as 37,000 acres (15,000 hectares). We will never know with certainty, but 8,000 hectares seems reasonable.
5. John Bartlett Brebner, *The Neutral Yankees of Nova Scotia: A Marginal Colony During the Revolutionary Years* (New York: Columbia University Press, 1937), 97.
6. Graeme Wynn, "Late Eighteenth Century Agriculture on the Bay of Fundy Marshlands," *Acadiensis* 8, 2 (1979): 80-89; the original work, an MA research paper submitted in 1969 to the Department of Geography, University of Toronto, was entitled "The Utilization of the Chignecto Marshlands of Nova Scotia and New Brunswick, 1750-1800." The text of this work was published many years later by the Tantramar Heritage Trust, with an extended

introductory reflection on the original research and work, by others, since: Graeme Wynn, "Of Time and Tides in the Settlement of Chignecto," *Culture and Agriculture on the Tantramar Marshes* (Sackville: Tantramar Heritage Trust, 2012), vii–xxiii.

7 Here most particularly perhaps E.P. Thompson, *The Making of the English Working Class* (New York: Pantheon Books, 1964), and the excitement generated by the emerging work on New England town life by such scholars as Kenneth Lockridge, Philip Greven, John Demos, and Michael Zuckerman, each of whom published notable monographs in 1970, as well as conversations with James T. (Jim) Lemon of the University of Toronto, Department of Geography, then working on *The Best Poor Man's Country: A Geographical Study of Early Southeastern Pennsylvania* (Baltimore: Johns Hopkins Press, 1972).

8 Clark (*Acadia*, vii) characterizes the Acadians who are at the heart of his book as "those dedicated, stubborn, resilient, pettifogging, inventive, exasperating, peace-loving, and in many ways altogether magnificent people."

9 Sherman Bleakney, *Sods, Soils and Spades: The Acadians of Grand Pré and their Dykeland Legacy* (Montreal and Kingston: McGill-Queen's University Press, 2004). Understanding of the reclamation, use, and ecologies of tidal marshlands, especially along the St. Lawrence River, has been greatly advanced by Matthew Hatvany, *Marshlands: Four Centuries of Environmental Change on the Shores of the St. Lawrence* (Quebec: Presses de l'Université Laval, 2003) and in other works listed in the bibliography of this volume.

10 Perhaps most markedly in the work of photographer Thaddeus Holownia. See *inter alia*: Peter Sanger, *Lightfield: The Photography of Thaddeus Holownia* (Kentville, NS: Gaspereau Press, 2018); Sarah Fillmore, Peter Sanger, David Diviney, and Thaddeus Holownia, *The Nature of Nature: The Photographs of Thaddeus Holownia, 1976-2016* (Halifax: Art Gallery of Nova Scotia, 2017). Thaddeus Holownia, *Tantramar Revisited, Revisited* (Jolicure, New Brunswick: Anchorage Press, 2018) and Thaddeus Holownia and Douglas Lochhead, *Dykelands* (Montreal and Kingston: McGill-Queen's University Press, 1989).

11 G.M. Grant, and L.R. O'Brien, *Picturesque Canada: The Country as It Was and Is* (Toronto: Belden Brothers, 1882).

12 Robert A. Mackinnon, "The Historical Geography of Agriculture in Nova Scotia, 1851-1951," PhD diss., University of British Columbia, 1991; Robert Summerby-Murray, "Interpreting Cultural Landscapes: A Historical Geography of Human Settlement on the Tantramar Marshes, New Brunswick," in *Themes and Issues in Canadian Geography III*, ed. Christoph Stadel (Salzburg: Salzburger Geographische Arbeiten, 1999), 157–74.

13 Attribution to High Marsh Road at catalogue entry also suggests that the image was reversed in printing. See https://www.mta.ca/marshland/topic1_environment/cunningham4b_29.htm.

14 Douglas Lochhead, "September 2," *High Marsh Road: Lines for a Diary* (Toronto: Anson-Cartwright, 1980, republished Fredericton: Goose Lane, 1996).

15 Paul R. Josephson, *Industrialized Nature: Brute Force Technology and the Transformation of the Natural World* (Washington, DC: Island Press/Shearwater, 2002).

16 James C. Scott, *Seeing Like a State: How Certain Schemes to Improve the Human Condition Have Failed* (New Haven: Yale University Press, 1998), 4.

17 Ronald Rudin, "The First Acadian Film: Visibility, Modernity, and Landscape in *Les aboiteaux*," *Canadian Historical Review* 96, 4 (December 2015): 507–33. On pages 528–31, Rudin notes that the English version of *Les aboiteaux* (a ten-minute film released as *The Dikes*) offers a bland rendition, stripped of the context and tensions in the original. The

NFB has available an English-subtitled version of the original. For LeBlanc, see Naomi E.S. Griffiths, *The Golden Age of Liberalism: A Portrait of Roméo LeBlanc* (Toronto: James Lorimer, 2011).

18 Luc Léger, "L'image du Parti acadien et de son projet autonomiste dans le journal L'Évangéline," *International Journal of Canadian Studies/Revue internationale d'études canadiennes* 45–46 (2012): 91–108. An account of these slightly later developments can be found in chapter 4 of Donald Savoie, *I'm from Bouctouche, Me: Roots Matter* (Montreal and Kingston: McGill-Queen's University Press, 2009).

19 Quotes from Rudin, "First Acadian Film," 519.

20 Irish writer and poet Patrick R. Chalmers, "Roundabouts and Swings," *Green Days and Blue Days* (1914).

21 H.D. Thoreau, "Walking," *The Atlantic Monthly, A Magazine of Literature, Art, and Politics* 9, 56 (June 1862): 657–74 and at Project Gutenberg: http://onlinebooks.library.upenn.edu/webbin/gutbook/lookup?num=1022; Raymond Williams, "Nature," *Keywords: A Vocabulary of Culture and Society* (London: Fontana, 1975).

22 Cicero cited by Colin M. Coates, "Back to the Land," in *The Nature of Canada*, ed. Colin M. Coates and Graeme Wynn (Vancouver: OnPoint Press, an imprint of UBC Press, 2019), 124.

23 Carl O. Sauer, "The Morphology of Landscape," (first published 1925 but most accessible) in *Land and Life; a Selection from the Writings of Carl Ortwin Sauer*, ed. John Leighly (Berkeley: University of California Press, 1967), 315–50.

24 For a brief discussion of gardens as third nature and a reproduction of the Frontispiece to l'Abbé de Vallemont, *Curiositez de la nature et de l'art* (1705), which portrays three natures, see Andrew Ray (Plinius), "Third Nature," June 5, 2009, at: https://some-landscapes.blogspot.com/2009/06/third-nature.html. W.A. Johnston, "Third Nature: The Co-evolution of Human Behavior, Culture, and Technology," *Nonlinear Dynamics, Psychology and Life Sciences* 9, 3 (July 2005): 235–80.

25 Bill McKibben, "The Emotional Core of *The End of Nature*," *Organization and Environment* 18, 2 (June 2005): 182–85.

26 There are of course many important contributions to this debate, from arguments that the very idea of nature was a human construction, see Bruno Latour, *We Have Never Been Modern* (Cambridge, MA: Harvard University Press, 1993) to Michael Pollen, *Second Nature: A Gardener's Education* (New York: Grove Press, 2008).

27 Stephen Jay Gould, *Time's Arrow: Time's Cycle: Myth and Metaphor in the Discovery of Geological Time* (Cambridge, MA: Harvard University Press, 1987), 15–16, 200.

28 American author Tom Wolfe had an unequivocal answer to this question, expressed in the title of the novel from which this quotation is taken: *You Can't Go Home Again* (New York: Harper and Bros., 1940), 602.

29 Charles G.D. Roberts, "Tantramar Revisited," *The Week* (December 1883), and in *Selected Poems of Sir Charles G.D. Roberts* (Toronto: Ryerson, 1936), 50-52, lines 11–14, 29–30, 51–52, 55, 58–59, 61–64.

30 Douglas Lochhead, "The Meeting," in *The Atlantic Anthology: Volume II, Poetry*, ed. Fred Cogswell (Charlottetown, PEI: Ragweed/ECW Press, 1985) and in Douglas Lochhead, *The Full Furnace: Collected Poems* (Toronto: McGraw-Hill Ryerson, 1975).

31 Douglas Lochhead, "So Close," *Fiddlehead* 143 (Spring 1985): 28.

32 Douglas Lochhead, "Poet Talking," *Tiger in the Skull: New and Selected Poems 1959-1985* (Fredericton: Fiddlehead Poetry Books/Goose Lane Editions, 1986).

33 See for some reflection on this, John Van Rys, "The Range of Douglas Lochhead," *Dalhousie Review* 67, 1 (1987): 133–37.
34 Holownia and Lochhead, *Dykelands*, 26; Douglas Lochhead, "Hard Statements" in *Tiger in the Skull.*
35 Douglas Lochhead, "Tracks," in *Atlantic Anthology: Vol. 2, Poetry*, ed. Fred Cogswell (Charlottetown, PEI: Ragweed/ECW, 1985).
36 The "mysteries" phrase is from Tony Tremblay, James W. Johnson, and Alexandra Cogswell, "Douglas Lochhead," New Brunswick Literature Curriculum in English, available online at: https://nblce.lib.unb.ca/resources/authors/douglas-lochhead.
37 Lochhead, "October 9," *High Marsh Road*. Rod Giblett, "The Marsh Lies Rich and Wanton: The Tantramar Marshes," in *Canadian Wetlands Places and People*, ed. Charles G.D. Roberts and Douglas Lochhead (Chicago: Intellect, 2014), 71–94. Ann Munton, "Return, Toronto to the Tantramar: Regional Poetics, the Long Poem and Douglas Lochhead," in *The Atlantic Anthology: Volume 3, Critical Essays*, ed. Terry Whalen (Charlottetown, PEI: Ragweed/ECW Press, 1985), 251–61.
38 Howell G.M. Edwards, "*Will-o'-the-Wisp:* An Ancient Mystery with Extremophile Origins?" *Philosophical Transactions of the Royal Society* 372, 2030 (2014), https://doi.org/10.1098/rsta.2014.0206.
39 Edward Relph, *Place and Placelessness* (London: Pion, 1976); quote from David Seamon, "A Singular Impact: Edward Relph's *Place and Placelessness*," *Environmental and Architectural Phenomenology Newsletter* 7, 3 (Fall 1996), 5–8.
40 *Inter alia* note the work of Thaddeus Holownia referenced in note 10 above and works by Alex Colville. Holownia and Colville both taught at Mount Allison University, Sackville, and its Fine Arts Program and the presence on campus of the Owens Art Gallery have contributed greatly to the production and display of marshland imagery. In a different register, see Amanda Dawn Christie, *Spectres of Shortwave,* documentary film: 113 minutes (Canada, 2016); Stephen Cass, "This Artist Made a Radio Out of a Kitchen Sink. Amanda Dawn Christie's Work Commemorates the Fading Glory of Shortwave Radio," *IEEE Spectrum*, September 27, 2018, at: https://spectrum.ieee.org/geek-life/profiles/this-artist-made-a-radio-out-of-a-kitchen-sink; Amanda Dawn Christie, "The Marshland Radio Plumbing Project" (Sackville, 2009) at: http://www.amandadawnchristie.ca/the-marshland-radio-plumbing-project-page; Marc Montgomery, "Radio Canada International: Spectres of Shortwave," https://www.rcinet.ca/en/2019/01/15/radio-canada-international-spectres-of-shortwave/. Also worth noting is Janet Crawford, *Matter of Time*, a ten-track album of songs with lyrics from Douglas Lochhead poems at https://www.deezer.com/us/album/1074713.
41 "inhabit" from Tremblay et al., "Douglas Lochhead,"; "human need" from Relph, *Place and Placelessness*, 147.
42 Matthew McClearn, "Will Sackville's Dikes Finally Fall? Rising Seas Could Ruin Land Acadians Turned from Marshes to Farms," *Globe and Mail*, April 2, 2018.
43 "Nova Scotia at Risk of Becoming an Island if Dikes Not Fixed: Officials," *Globe and Mail*, November 16, 2016.
44 Cultural landscapes by UNESCO definition "are illustrative of the evolution of human society and settlement over time, under the influence of the physical constraints and/or opportunities presented by their natural environment and of successive social, economic and cultural forces, both external and internal" (UNESCO 2005). Guidelines on the inscription of specific types of properties on the world heritage list are available at: http://whc.unesco.org/archive/opguide05-annex3-en.pdf.

45 Kate Sherren, Logan Loik, and James A. Debner, "Climate Adaptation in 'New World' Cultural Landscapes: The Case of Bay of Fundy Agricultural Dykelands (Nova Scotia, Canada)," *Land Use Policy* 51 (February 2016): 267–80.
46 Robert Z. Melnick, "Deciphering Cultural Landscape Heritage in the Time of Climate Change," *Landscape Journal: Design, Planning, and Management of the Land* 35, 2 (2016): 287–302.
47 "countless dialogues" from Anne W. Spirn, *The Language of Landscape* (New Haven: Yale University Press, 1988), 40; "jumble" from Marion Shoard, "Why Landscapes Are Harder to Protect than Buildings," in *Our Past Before Us: Why Do We Save It?*, ed. David Lowenthal and Marcus Binney (London: Temple Smith, 1981), 83–108 quote from 91.
48 Aiden Fudge, "Memory, Place and Change: A Landscape Narrative of the Tantramar Marshes," Master of Arts thesis, University of Guelph, 2019, brings this approach to bear on the Fundy marshlands. Many of the numerous illustrations in this document are striking and excellent; the landscape narrative is, unfortunately, less impressive.
49 Douglas Lochhead, "A Loud Place," *Fiddlehead* 143 (Spring 1985): 28.
50 See L. Scazzosi, "Reading and Assessing the Landscape as Cultural and Historical Heritage," *Landscape Research* 29, 2 (2004): 335–55.

Preface: In Search of the MMRA

1 According to the 2016 census, the Moncton metropolitan area has a population of roughly 150,000, third in Atlantic Canada after Halifax and St John's.
2 Davidson, "How Much Wetland Has the World Lost?" 934–41. Regarding the decline of salt marsh in the Bay of Fundy region, see Thurston, *Tidal Life*, 86; and Percy, "Dykes, Dams, and Dynamos," 6. For more on the Ramsar Convention, visit https://www.ramsar.org.
3 Nicholas, "Putting Wetlands in Perspective," 31; Hatvany, "Wetlands and Reclamation," 14: 273.
4 The various components of the wetlands literature are well represented in a number of anthologies: Williams, *Wetlands;* Mitsch and Gosselink, *Wetlands;* and Tiner, *Tidal Wetlands Primer.*
5 A case in point is Rod Giblett's *Canadian Wetlands,* which discusses the Bay of Fundy marshlands at length, without ever referring to the MMRA.
6 Thurston, *Tidal Life,* 87; also see Thurston, *A Place between the Tides.*
7 Nova Scotia, Department of Agriculture and Marketing, *Maritime Dykelands,* vii.
8 Ibid., 90.
9 Since my original discovery of the Nova Scotia records, they have been removed from that basement location, with some being catalogued into a database. As for the New Brunswick records, Mike Green, who works for the provincial Department of Agriculture, says that he transported the boxes from Amherst to Moncton in the early 1980s.
10 This request ultimately provided over 1,500 pages of documents, but I would not have received that material without the assistance of Catherine Viveiros, in the Office of the Information Commissioner, who handled my complaint when Fisheries and Oceans Canada did not meet its statutory requirements. I was encouraged informally by Suzanne Legault (then the Information Commissioner), in a chance conversation at a conference, to file the complaint if I felt that the statute had not been observed. So I encourage researchers both to use Access to Information and Privacy (ATIP) requests as a tool, and to insist on their rights.

11 Marsh bodies also existed in New Brunswick, at least until the 1960s, when provincial legislation effectively put them out of business, a matter discussed at length in Chapter 2. When the New Brunswick marsh bodies were forced to close down, there was no one left to take responsibility for their documents, and so I was unable to find any of their minute books.

A Note on Sources

1 Library and Archives Canada (hereafter LAC), RG 124, series A3, file 165-M5, vols. 25, 26, 132.

Prologue

1 For an extended discussion of the film, see my article "The First Acadian Film." An English-subtitled version of the film (*The Dikes*) can be found at https://www.nfb.ca/film/dikes_en/.
2 There is a vast literature dealing with the role of Acadians in constructing the dykes and aboiteaux. It is referenced at length in Chapter 1, but for an introduction to the subject, see Bleakney, *Sods, Soils and Spades;* Kennedy, *Something of a Peasant Paradise?;* Cormier, *Les aboiteaux en Acadie;* Hatvany, "The Origins of the Acadian Aboiteau."
3 The post-deportation management of the dykelands is discussed in such works as Wynn, "Late Eighteenth-Century Agriculture," and Summerby-Murray, "Interpreting Cultural Landscapes."
4 Letter from Sackville and Westmorland Agricultural Society to J.G. Gardiner, July 7, 1939, LAC, RG 17, file 559-17; 3295 file part 1.
5 Interview with Ernie Partridge, Dorchester, NB, September 2, 2015.
6 I discuss the significance of the bicentenary to the emergence of "modernity" in Acadie in *Remembering and Forgetting in Acadie*, 193–212.
7 Using mother tongue as an imperfect guide, there were 234,753 Acadians in 1951 and 261,206 ten years later.
8 There is a vast literature on the emergence of modernity in postwar Acadie. For an excellent introduction, see Belliveau, *In the Spirit of '68.*
9 Rudin, *Remembering and Forgetting in Acadie,* Part I. The different stories about the past evoked by this landscape in 2004 are also presented in a film that I produced and which was directed by Leo Aristimuño: *Life after Île-Croix* (Montreal: National Film Board, 2006), https://www.nfb.ca/film/life-after-ile-ste-croix.
10 Throughout the text, I have used the French form (Grand-Pré), except when referring to the contemporary municipality or the marsh body that looks after dykeland there. In those cases, I follow local practice and use the English form, Grand Pre.
11 Rudin, *Kouchibouguac*. Individuals removed from their lands at Kouchibouguac tell their stories on the website that I created in collaboration with Philip Lichti: http://returningthevoices.ca.
12 In this regard, Yves Cormier wrote about how the construction of the protective structures had moulded a certain Acadian identity: "Because they had to work together as a group, the Acadians developed a sense of community, as well as a tendency to be sedentary" (*Les aboiteaux en Acadie,* 73). I have translated this and all subsequent French passages in the book into English.

13 The list of Acadian cultural productions inspired by the marshland environment is long. To take only one example, it figured prominently in Calixte Duguay's popular song, *Les aboiteaux* (1976), which mournfully speaks to the scattering of a diasporic population, with the aboiteaux marking the landscape where they "are waiting for the land around them to reawaken" (Les aboiteaux attendent quelque part que le pays s'éveille). Twenty years later, at the first Congrès mondial acadien, a gathering of the dispersed Acadian clan staged in the same region of New Brunswick where *Les aboiteaux* had been filmed, Duguay spoke about how (like many Acadian artists) he had left Acadie to work in Montreal, but had not felt welcome on returning home. Using the aboiteau metaphor, he described how it had been possible for him to leave "through the opening in the aboiteau to go out and see the world," but on returning he discovered "un clapet fermé." In response, Michel Doucet – a prominent leader from New Brunswick – picked up on the role of *le clapet*. He recognized that Acadians often left New Brunswick for opportunities elsewhere, and so it was necessary to "to open the gates (clapets) of its aboiteaux to allow those who want to leave to go out and find the inspiration they require." The problem came when they wanted to return, because salt water (or in this case people) was not supposed to flow back in, with the result that the expatriates often found that "le clapet was closed to prevent a destructive flood." Expressing resentment of those who had not stayed at home to improve their society, Doucet pointed out: "The aboiteau would serve no function if the gates were always left open." (Duguay, "L'Acadie dans le village global," in *Le Congrès mondial acadien*, 83; Doucet, "Rapport général des conférences," *Le Congrès mondial acadien*, 620). The quotes above were originally in French.
14 UNESCO, Citation declaring the Landscape of Grand-Pré as a World Heritage Site: http://whc.unesco.org/en/list/1404. For more on Grand-Pré as a UNESCO World Heritage Site, see Campbell, *Nature, Place, and Story*, ch. 2.
15 Léonard Forest, Présentation of *Les aboiteaux*, June 29, 1954, National Film Board of Canada (NFB) Archives, file 53-144. The quote was originally in French.
16 This particular representation was created in the context of a film that required the involvement of the MMRA, which was assured by the NFB that it would provide "favourable publicity." Michael Spencer to Ralph MacKay, July 15, 1954, NFB Archives, file 53-144.
17 *Summary Report on Project NB 22*, 45–46.
18 Scott, *Seeing Like a State*. Scott's work has been a lightning rod, attracting the attention of researchers in a number of disciplines. In addition to the works discussed below, a reader can get a sense of the debates that Scott has stimulated by looking at the exchanges over *Seeing Like a State*, in *The Good Society* 10 (2001): 36–51.
19 Scott, *Seeing Like a State*, 313.
20 Ibid., 318.
21 Ibid.
22 Ibid., 332.
23 Ibid., 4, 94.
24 Scott, "High Modernist Social Engineering," 8.
25 Zahra, "Lost Children," 47.
26 Woods, *Rehabilitation*, 10; Dummitt, *The Manly Modern*, 105.
27 Fiege, "The Weedy West," 24.
28 Fiege, *Irrigated Eden*, 207.
29 Stunden Bower, *Wet Prairie*, 168.

30 Scott, "High Modernist Social Engineering," 19–24.
31 Macfarlane, *Negotiating a River*, 139, 142.
32 Kenny and Secord, "Engineering Modernity," 26. The same focus on rehabilitation was also evident in the creation of Kouchibouguac National Park, a project that was designed not only to transform the landscape that had long supported hundreds of families but also to "fix" the residents. These individuals were so poor that government officials determined that they needed to be uprooted and their way of life changed through such means as providing courses to women so that they could do a better job of raising their children. I discuss this in *Kouchibouguac*, 106–13.
33 Stunden Bower, *Wet Prairie*, 78.
34 Pisani, *Water and American Government*, 292, 294.
35 Ibid.
36 Loo, "High Modernism," 44–46.
37 Armstrong, Evenden, and Nelles, *The River Returns*, 177–78. While *The River Returns* deals with the PFRA in the context of a single river, there is no well-researched monograph devoted to this important agency. James Gray's *Men against the Desert* is valuable for its use of interviews with major players in the history of the PFRA, but if he consulted archives, he did not indicate that he had, in what was, to be fair, a work for a popular audience. The title for this book was inspired by the title of Gray's.
38 Gilbert, "Low Modernism and the Agrarian New Deal," 131. Gilbert's use of the term "low modernism" was designed to set himself apart from Scott. Gilbert discusses the concept at length in his *Planning Democracy*.
39 Loo, "High Modernism," 38.
40 Loo, *Moved by the State*, 197–98.
41 Scott, *Seeing Like a State*, 332.
42 Hatvany, *Marshlands*, 24–25.
43 Cronon, *Nature's Metropolis*, xix. The same term is usefully employed by Richard Judd, who found that it was often a "'second nature' as opposed to the more pristine 'first nature' landscapes that served as the beginning point for most early environmental histories." *Second Nature*, 13.
44 Aubert Hamel, "La récupération et la mise en valeur des alluvions maritimes du Saint-Laurent," *L'Agriculture* 20, 3 (1963): 77–83, cited in Hatvany, *Marshlands*, 123.
45 Interview with Ernie Partridge, Dorchester, NB, June 11, 2018.
46 William Cronon, "Foreword," in Langston, *Where Water and Land Meet*, xii.
47 Worster, *Rivers of Empire*, 7.
48 Reisner, *Cadillac Desert*, 481, 485. Reisner's view of the Bureau of Reclamation was at odds with that of Donald Pisani, noted above.
49 Matas, "N.B. River Leads Endangered List."
50 Teal and Teal, *Life and Death of the Salt Marsh*, 217.
51 Loo, "High Modernism," 49.
52 Fiege, *Irrigated Eden*, 208.

Chapter 1: Out to Sea

1 Kindervater, *Flooding Events in Nova Scotia*.
2 Archibald to Deputy Minister of Agriculture, September 17, 1943, LAC, RG 17, 3295, pt. 1.

3   Ibid.
4   Nova Scotia Archives, Royal Commission on Provincial Development and Rehabilitation Fonds, book 1, vol. 3, Hearings in Middleton, N.S., July 29, 1943.
5   Ibid.
6   Report Submitted by the Dykeland Rehabilitation Committee on Urgent Repair Work Necessary for the Preservation of the 1944 Crops, November 1943, LAC, RG 17, 3295, pt. 1.
7   In the pages that follow, I am able to describe the day-to-day management of the drained marshlands only because I had access to a number of the minute books generated by marsh bodies. With the exception of the minutes of the Grand Pre Marsh Body, there are no records in archival collections, and so I have been dependent on the Nova Scotia records that I was fortunate to discover in private collections. The marsh bodies whose records have survived (along with the individuals who possess them) are identified in the Preface. The absence of New Brunswick records is discussed in Chapter 2.
8   Report Submitted by the Dykeland Rehabilitation Committee, LAC, RG 17, 3295, pt. 1.
9   Nova Scotia Archives, Royal Commission on Provincial Development and Rehabilitation Fonds, book 2, vol. 6, Hearings in Amherst, N.S., August 10, 1943.
10 Bleakney, *Sods, Soils and Spades*, 5.
11 Champlain, "Voyages of Samuel de Champlain," 1: 258.
12 Matthiessen, "Tidemarshes," xii.
13 Wicken, "Re-examining Mi'kmaq-Acadian Relations," 100; Kennedy, *Something of a Peasant Paradise*, 23.
14 Champlain, "Voyages of Samuel de Champlain," 1: 371.
15 Hatvany, "Wetlands and Reclamation," 277; Hatvany, Cayer, and Parent, "Interpreting Salt Marsh Dynamics," 1045.
16 Vileisis, *Discovering the Unknown Landscape*, 36.
17 Kennedy, *Something of a Peasant Paradise*, 24.
18 Nomination Grand Pré, *World Heritage Nomination Proposal*, 33.
19 Cormier, "Nos aboiteaux," 14. Even today, farmers sometimes have to travel several kilometres from their farm buildings on the uplands to their drained marshland.
20 Boardman, "Tidal Lands and Diked Marshes," 41. In the original, the word was exceptionally spelled "dike," not "dyke." I have standardized the spelling throughout the text.
21 Wynn, "Late Eighteenth-Century Agriculture," 80. He estimated that crops were covering between 13,000 and 20,000 acres at mid-century.
22 Kennedy, "Marshland Colonization," 44–45.
23 Clark, *Acadia*, 161. The names of the rivers here are the modern ones, imposed by English-speakers after the removal of the Acadians, who had previously imposed their own names in place of those used by the Mi'kmaq First Nation.
24 Daigle, "Acadian Marshland Settlement," pl. 29.
25 Boardman, "Tidal Lands and Diked Marshes," 36.
26 Interviews with Dick Haliburton, Avonport, NS, July 2017; George Trueman, Aulac, NB, July 2017.
27 Kennedy, *Something of a Peasant Paradise*, 26.
28 Roméo LeBlanc, *Travail de recherche*, 1954, 12, NFB Archives, file 53-144.
29 Bleakney, *Sods, Soils and Spades*, 54.
30 Ibid., 170.
31 Interview with Ernie Partridge, Dorchester, NB, September 2, 2015.
32 Bleakney, *Sods, Soils and Spades*, 45.

33 Ibid., 47, 37.
34 Ibid., 47–48.
35 Ibid., 46.
36 Ibid., 37.
37 Ibid., 183–84.
38 Ibid.
39 Nomination Grand Pré, *World Heritage Nomination Proposal*, 38. Ronnie-Gilles LeBlanc found a similar use of a lottery to distribute drained marshland at Memramcook, "Documents acadiens sur les aboiteaux," 44.
40 Ibid., 45.
41 Marshland Files, Memramcook West Marsh Body, Map 21.1, 1951, New Brunswick Department of Agriculture (hereafter NBDA).
42 As we saw in the Prologue, Shannon Stunden Bower has described a similar situation in terms of water management in the Canadian prairies, where individualism could conflict with the need for communal action when water moved across property lines. *Wet Prairie*, 168.
43 Kennedy, *Something of a Peasant Paradise*, 99–100.
44 Faragher, *A Great and Noble Scheme*, 333.
45 Cited in Nomination Grand Pré, *World Heritage Nomination Proposal*, 41.
46 Another family that came to this part of Nova Scotia at roughly the same time as the Palmeters was that of J.F.W. DesBarres, which took over the land of the Landry family in 1763. DesBarres had been General James Wolfe's aide-de-camp during the Seven Years' War, and subsequently established an estate known as Castle Frank along the Avon River. Interview with Jim Bremner, Falmouth, NS, July 12, 2017.
47 Nomination Grand Pré, *World Heritage Nomination Proposal*, 48; Desplanque and Mossman, "Storm Tides of the Bay of Fundy," 23–33.
48 Sebold, "*The Low Green Prairies of the Sea*," 121.
49 Wynn, "Late Eighteenth-Century Agriculture," 83–84.
50 Faragher, *A Great and Noble Scheme*, 419.
51 Wynn, "Late Eighteenth-Century Agriculture," 85.
52 I repeatedly heard the word "parmang" in interviews with Ernie Partridge. In the House of Commons debates in 1948 over the bill establishing the MMRA, Percy Black, an MP from Amherst, referred to a "word used by the old Acadians; it sounded like 'pamang.' I am afraid I do not know how to spell this word. It refers to the method of using sod in building the face of the dikes" (Canada, Parliament, House of Commons, *Debates*, 20th Parl., 4th Sess., 5 [1947–48]: 4562). As for the French form of the word, in documentation prepared for *Les aboiteaux*, Roméo LeBlanc referred to *un parment* (Roméo LeBlanc, *Travail de recherche*, 1954, 18, NFB Archives, file 53-144). For his part, Adolphe Le Blanc, a *sourd de marais* from Memramcook, referred to "les parements" (Interview with Père Anselme Chiasson, late 1950s, *Les cahiers de la Société historique acadienne* 19 [1988]: 52).
53 Wynn, "Late Eighteenth-Century Agriculture," 87.
54 Faragher, *A Great and Noble Scheme*, 418; *Nova Scotia Statutes*, 1760, c. 7.
55 *Nova Scotia Statutes*, 1760, c. 7.
56 Ibid.
57 After 1783, with the creation of New Brunswick, there would be two sets of legislation that were generally identical to one another. To be sure, this exercise in local self-government was not always egalitarian, as is to be expected given the nature of democracy in the long

nineteenth century, during which the majority of Canadians were denied the right to vote or hold office for reasons connected to their gender, wealth, or identity. In the case of the marsh organizations, wealthier proprietors were favoured by aspects of the governing legislation that generally required decisions to be approved by owners holding a majority of the land, opening the door to a minority of owners dictating policy. (See, for instance, *Nova Scotia Statutes*, 1900, c. 12, art. 4.) In fact, however, the marsh body minute books that I consulted rarely indicated a vote taking place; and while there may have been a tendency for deference toward wealthier proprietors, I found little evidence of significant debate over how to manage the drained marshes.

58 An identical plaque was erected on the same day at Port Williams, only a few kilometres from Grand-Pré, to mark the anniversary of the Bishop Beckwith Marsh Body.
59 Wynn, "Late Eighteenth-Century Agriculture," 87–89.
60 Warren, *Tidal Marshes and Their Reclamation*, 97.
61 Boardman, "Tidal Lands and Diked Marshes," 35. The development of the hay economy is also discussed in Summerby-Murray, "Commissioners of Sewers."
62 Warren, *Tidal Marshes and Their Reclamation*, 97, 99.
63 Hatvany, "Wetlands and Reclamation," 277.
64 Warren, *Tidal Marshes and Their Reclamation*, 86.
65 Baird, *Report on Dyke Reclamation*, 2.
66 Warren, *Tidal Marshes and Their Reclamation*, 97; Baird, *Report on Dyke Reclamation*, 2.
67 Supplementary Report on Beef Cattle Industry in Westmorland County, N.B. and Cumberland County, N.S, LAC, RG 17, 3295, pt. 1 (emphasis in original). The Maritime Stock Breeders Association had a standing Beef Cattle Committee. Faced with the crisis in the marsh areas on either side of the Nova Scotia–New Brunswick border, the federal government created a special subcommittee to suggest solutions. The members included W.W. Baird and C.F. Bailey (superintendents of the experimental farms at Nappan and Fredericton); S.F. Wood and C.E. Boulden (livestock superintendents for New Brunswick and Nova Scotia); and J.W. Graham, Production and Market Services, Canada, Department of Agriculture.
68 Baird, *Report on Dyke Reclamation*, 2.
69 Urquhart, *Historical Statistics of Canada*, 367; Sandwell, chs. 3–4.
70 Report of a Survey of the Sub-Committee of the Maritime Beef Cattle Committee on Conditions Affecting the Beef Cattle Industry in the Chignecto District of New Brunswick and Nova Scotia, February 1940, LAC, RG 17, 3295, pt. 1.
71 Baird, *Report on Dyke Reclamation*, 2–3.
72 Report of a Survey of the Sub-Committee of the Maritime Beef Cattle Committee, LAC, RG 17, 3295, pt. 1. Focusing on purchase prices in the Annapolis Valley, Gordon Haase found that they had stayed at around $100 per acre through the first three decades of the twentieth century, before falling to $36 by the 1930s and $31 by the 1940s. Haase, *Some Economic Aspects*, 18.
73 Minutes of Meeting ... Convened to Consider Problems Relating to Marshland Rehabilitation, September 1943, LAC, RG 17, 3295, pt. 1.
74 Supplementary Report on Beef Cattle Industry, LAC, RG 17, 3295, pt. 1.
75 Lattimer, *A Study of 128 Profitable Cumberland County Farms*.
76 Nova Scotia Department of Agriculture (hereafter NSDA), Agricultural Engineering Division, *Nova Scotia Marsh Survey*, 1939. There was also a much more modest study carried

out in 1939 by H.R. Hare of the federal Department of Agriculture's Economics Division. Tantramar Marsh Survey, 1939, LAC, RG 17, vol. 3660, file N-6-56.
77 NSDA, *Nova Scotia Marsh Survey*, 117.
78 Report of a Survey of the Sub-Committee of the Maritime Beef Cattle Committee, LAC, RG 17, 3295, pt. 1.
79 NSDA, *Nova Scotia Marsh Survey*, 15.
80 Report of a Survey of the Sub-Committee of the Maritime Beef Cattle Committee, LAC, RG 17, 3295, pt. 1.
81 NSDA, *Nova Scotia Marsh Survey*, 89–90 (emphasis in original).
82 Report of a Survey of the Sub-Committee of the Maritime Beef Cattle Committee, LAC, RG 17, 3295, pt. 1.
83 Tantramar Marsh Survey, LAC, RG 17, vol. 3660, file N-6-56.
84 Léonard Forest, *The Dikes* (English version of *Les aboiteaux*), June 29, 1954, National Film Board of Canada (NFB) Archives, file 53-144.
85 Hatvany, *Marshlands*, 119–20. I have no evidence that the Kamouraska experiment was known to the farmers at Grand-Pré.
86 MMRA report on Grand Pre Marsh Body, September 10, 1949, Acadia University Archives, MMRA Fonds, 1998.001/1/2. The Palmeter family emerges in these pages from time to time in terms of dykeland management at Grand-Pré and across the region.
87 Haase and Packman, *Marshland Utilization*, 15–17. This is one of the few studies that allows careful analysis of how marshland farms operated. Census material is generally useless as the data are aggregated for areas that included both farms that had drained marsh and those that did not.
88 Ibid., 31.
89 Tantramar Marsh Survey, LAC, RG 17, vol. 3660, file N-6-56.
90 Report of a Survey of the Sub-Committee of the Maritime Beef Cattle Committee, LAC, RG 17, 3295, pt. 1.
91 MMRA report on Grand Pre Marsh Body, September 10, 1949, Acadia University Archives, MMRA Fonds, 1998.001/1/2. Hatvany also reported problems with the Kamouraska aboiteaux constructed with heavy machinery, although it is not clear whether the use of that machinery was the cause. In any event, once damaged, those aboiteaux were not repaired for decades. Hatvany, *Marshlands*, 120.
92 MMRA report on Grand Pre Marsh Body, September 10, 1949, Acadia University Archives, MMRA Fonds, 1998.001/1/2.
93 NSDA, *Nova Scotia Marsh Survey*, 38–39.
94 Ibid., 17, 120, 141.
95 Ganong, "The Vegetation of the Bay of Fundy Salt and Diked Marshes," 178.
96 Desplanque and Mossman, "Tides and Their Seminal Impact," 293.
97 Sheldon, *To Canada*, 24.
98 *Fifty Years of Progress on Dominion Experimental Farms*, 103. Tiding at Nappan was similarly described positively in the farm's 1889 report, which noted: "The high tides of December last that broke the dykes and flooded the marsh were a benefit to those parts of the marsh that were well drained, as the salt water ran off quickly but left a deposit of new mud." Cited in Calder, *100 Years of Agricultural Research*, 3.
99 Falmouth Village Great Dyke Marsh Body, Minutes, July 16, 1906; Warren, *Tidal Marshes and Their Reclamation*, 95.

100 Thurston, *A Place between the Tides*, 131–32.
101 The terminology changed over time, but the practice figured in the powers of commissioners of sewers until introduction of a new regime with creation of the MMRA; there was no reference to tiding – for the first time – in the New Brunswick and Nova Scotia legislation of 1949 that set the conditions for marsh bodies to receive aid from the MMRA. That new legislation is discussed below.
102 NSDA, *Nova Scotia Marsh Survey*, 142; Appendix, 127–28.
103 Ibid., 100; Appendix, 51.
104 Ibid.
105 Anstey, *One Hundred Harvests*, 77. The Ontario Agricultural College described Archibald as having "strengthened the application of scientific research to agricultural problems" (https://www.uoguelph.ca/oac/140faces/edgar-spinney-archibald). Even though he played an important role in the history of Canadian agriculture during the first half of the twentieth century, Archibald has been given little attention by historians, perhaps because his papers are not easily found at Library and Archives Canada. There are no easily accessible fonds regarding the experimental farms, and so to understand Archibald's role in the various portfolios with which he was involved requires extracting his correspondence from files dealing with specific issues. I was therefore introduced to Archibald through the LAC files pertinent to marshlands during the 1940s and early 1950s.
106 James Gray described Archibald as being more important to the creation of the PFRA "than all the others put together." He "spearheaded the campaign to bring the PFRA into existence and was the driving force that eventually got it functioning." Gray, *Men against the Desert*, 100.
107 Archibald to G.S.H. Barton, September 4, 1940, LAC, RG 17, 3295, pt. 1.
108 Report of W.D. Davies, Assistant Chief, Department of Agriculture, Production Services, September 13, 1940, LAC, RG 17, 3295, pt. 1.
109 Archibald to G.S.H. Barton, September 26, 1940, LAC, RG 17, 3295, pt. 1.
110 Longfellow, "Evangeline: A Tale of Acadie (1847)," in *Poems and Other Writings*, 58. Quite aside from its reference to tiding, Longfellow's poem had a significant impact on the formation of Acadian identity for nearly a century. Acadians internalized the heroic qualities of the fictitious Evangeline, who was separated from her boyfriend at the time of the deportation, but stoically travelled across North America to find him. Acadian leaders tried to turn such stoicism into a national trait. I have written about the power of the poem in both *Remembering and Forgetting in Acadie*, ch. 5, and *Kouchibouguac*, ch. 5.
111 Archibald to G.S.H. Barton, September 26, 1940, LAC, RG 17, 3295, pt. 1.
112 Ganong, "The Vegetation of the Bay of Fundy Salt and Diked Marshes," 178.
113 NSDA, *Nova Scotia Marsh Survey*, Appendix, 129; Archibald to G.S.H. Barton, September 26, 1940, LAC, RG 17, 3295, pt. 1.
114 NSDA, *Nova Scotia Marsh Survey*, 39.
115 Archibald to G.S.H. Barton, February 14, 1942, LAC, RG 17, 3295, pt. 1. The idea of blocking rivers with aboiteaux would later evolve into the construction of tidal dams, a subject that is the focus of Chapter 3.
116 Archibald to G.S.H. Barton, June 4, 1942, LAC, RG 17, 3295, pt. 1.
117 Scott, *Seeing Like a State*, 342–43, 336.
118 Gray, *Men against the Desert*, 103. Russell's role in the PFRA is described in Spence, *Survival of a Vision*. Spence enthusiastically shows how Russell in late 1943, only months after his

119 Archibald to John A. McDonald (Nova Scotia Minister of Agriculture) and A.C. Taylor (New Brunswick Minister of Agriculture), August 26, 1943, LAC, RG 17, 3295, pt. 1.
120 B. Russell, "Marsh Lands of the Maritime Provinces, Report of Reconnaissance," November 22, 1943, LAC, RG 17, 3295, pt. 1.
121 Ibid.
122 Ibid.
123 Ibid.
124 Tantramar Marsh Survey, LAC, RG 17, vol. 3660, file N-6-56; Archibald, "Maritime Dykeland Rehabilitation," 202.
125 Gardiner to A.C. Taylor, May 7, 1943, LAC, RG 17, 3295, pt. 1.
126 E.S. Hopkins (Associate Director, Central Farm) to G.S.H. Barton, August 11, 1943, LAC, RG 17, 3295, pt. 1.
127 Nova Scotia, House of Assembly, Resolution, April 6, 1943, LAC, RG 17, 3295, pt. 1.
128 Canada, House of Commons, Special Committee on Reconstruction and Re-establishment, Hearings, April 8, 1943.
129 Ibid.
130 In the fall of 1943, hearings of the New Brunswick Committee on Reconstruction also touched on the rehabilitation of dykelands. Summary notes from those hearings are housed in the New Brunswick Legislative Library in Fredericton (*Report of the New Brunswick Committee on Reconstruction*).
131 Nova Scotia Archives, Royal Commission on Provincial Development and Rehabilitation Fonds, book 1, vol. 4, Hearings in Kentville, N.S., July 30, 1943.
132 Archibald to Deputy Minister of Agriculture, September 17, 1943, LAC, RG 17, 3295, pt. 1.
133 The PFRA was represented by Ben Russell, who was already in the region to do research for his report discussed earlier, and by L.B. Thomson, superintendent of the experimental station in Swift Current, Saskatchewan. James Gray described Thomson as "Archibald's deputy and field-working alter ego" in connection with the PFRA. Gray, *Men against the Desert*, 218.
134 Archibald to Deputy Minister of Agriculture, September 17, 1943, LAC, RG 17, 3295, pt. 1.
135 Similarly, those with the most at risk were excluded from discussions in 1969 regarding the future of commercial fishing, which was scheduled to end following the creation of New Brunswick's Kouchibouguac National Park. As with this meeting about the marshlands, the most important discussions about fishing in the park took place among bureaucrats, with the fishers joining in only after a decision had already been made. Rudin, *Kouchibouguac*, 63–66.
136 Minutes of Meeting ... Convened to Consider Problems Relating to Marshland Rehabilitation, September 1943, LAC, RG 17, 3295, pt. 1.
137 Ibid.
138 Ibid.
139 Experimental Farm, Nappan, NS, *75th Anniversary*, n.p.
140 Baird, *Dominion Experimental Farm, Nappan, N.S.*, 33.
141 Barton to Archibald, August 31, 1949, LAC, RG 17, vol. 2836, file 1-17-1.
142 It is always difficult to prove a negative, but I found no reference to the farm at Nappan working with the farmers prior to 1943. Perhaps thinking that this was a provincial jurisdiction,

the Nappan officials might have communicated directly with their provincial counterparts, but I found no evidence of such communication.
143 Baird, *Report on Dyke Reclamation,* 9. In one of its few critical comments, *Maritime Dykelands* provides evidence from interviews with J.A. Roberts (who supervised the New Brunswick students) and J.W. Byers (who did the same for the Nova Scotia crew). As the latter put it: "'A couple of the students were interested in the work, the rest just had fun.' The project was 'a waste while dykes broke around the Bay.'" Nova Scotia, Department of Agriculture and Marketing, *Maritime Dykelands,* 65.
144 J.A. Roberts, "Report on Urgent Dyke repairs needed in Albert Co, NB," LAC, RG 19, vol. 432, file pt. 1, file 105-16.
145 Ibid.
146 Canada, *Public Accounts for the Fiscal Year Ended March 31, 1943,* A-6.
147 Ilsley to J.A. McDonald, December 29, 1942, LAC, RG 19, vol. 432, file pt. 1, file 105-16.
148 Archibald to Ilsley, December 1, 1943, LAC, RG 17 3295, pt. 1.
149 Archibald to A.M. Shaw (Chair, Agricultural Supplies Board), November 23, 1943; Report Submitted by the Dykeland Rehabilitation Committee on urgent repair work necessary for the preservation of the 1944 crops, LAC, RG 17, 3705, file pt. 242.
150 Minister of Agriculture to Governor-General-in-Council, May 2, 1944, LAC, RG 17, 3295, pt. 1.
151 PC 3821, April 22, 1944, LAC, RG 17, 3295, pt. 1.
152 Ilsley to J.A. McDonald, March 12, 1945, LAC, RG 19, vol. 432, file pt. 1, file 105-16.
153 Dentiballis Marsh Body, Minutes, September 2, 1943; September 2, 1944.
154 Archibald to Deputy Minister of Agriculture, January 15, 1945, LAC, RG 17, 3296, file 559-17.
155 J.W. Byers, "Estimate of Crop Saved by Dyke Repairs," November 1, 1944, LAC, RG 19, vol. 432, file pt. 1, file 105-16.
156 Archibald to Deputy Minister of Agriculture, March 12, 1945, LAC, RG 17, 3296, file 559-17.
157 Ilsley to A.S. MacMillan (Premier of Nova Scotia), July 25, 1945, LAC, RG 19, vol. 432, file pt. 1, file 105-16.
158 J.W. Byers, "Estimate of Crop Saved by Dyke Repairs," November 1, 1944, LAC, RG 19, vol. 432, file pt. 1, file 105-16.
159 Nova Scotia, Department of Agriculture and Marketing, *Maritime Dykelands,* 66.
160 Nappan Experimental Farm, Daily Journal, May 10-12, 1944. I owe a special thanks to John Duynisveld, a research biologist at the farm, who brought the existence of this journal (hidden in a basement cupboard) to my attention.
161 Ibid., November 4, 1944; August 1945; J.A. Roberts, "Report on Urgent Dyke repairs needed in Albert Co, NB," LAC, RG 19, vol. 432, file pt. 1, file 105-16. The situation at Harvey Bank is also discussed above at greater length.
162 Baird to Archibald, February 11, 1945, LAC, RG 17, 3296, file 559-17.
163 J.W. Byers, "Estimate of Crop Saved by Dyke Repairs," November 1, 1944, LAC, RG 19, vol. 432, file pt. 1, file 105-16.
164 W.W. Baird, "Annual Report on Maritime Marshland Work for the Year 1944," LAC, RG 19, vol. 432, file pt. 1, file 105-16.
165 Archibald to Deputy Minister of Agriculture, March 12, 1945, LAC, RG 17, 3296, file 559-17.

166 Ilsley to J.A. McDonald, March 12, 1945, LAC, RG 19, volume 432, file pt. 1, file 105-16.
167 In the following list, the contribution from Ottawa follows in parentheses after the pertinent order-in-council: PC 2919, April 24, 1945 ($43,000); PC 2/3368, August 8, 1946 ($60,000); PC 4549, November 12, 1947 (Nova Scotia, $60,000; New Brunswick, $40,000); PC 294, January 25, 1949 (Nova Scotia, $60,000; New Brunswick, $40,000).
168 Baird, *Report on Dyke Reclamation*, 11; Gardiner to Ilsley, August 8, 1945, LAC, RG 19, vol. 432, file pt. 1, file 105-16.
169 Archibald to Deputy Minister of Agriculture, July 19, 1945, LAC, RG 17, 3296, file 559-17.
170 Canada, *Report of the Minister of Agriculture for the Year Ended March 31, 1945*, 109.
171 Logan to A.W. Mackenzie, June 3, 1946, LAC, RG 19-F-2, vol. 4427, file 9320-02, pt. 1.
172 Mackenzie to Gardiner, June 5, 1946, LAC, RG 19-F-2, vol. 4427, file 9320-02, pt. 1.
173 Logan to Ilsley, June 29, 1946, LAC, RG 19, vol. 432, file pt. 1, files 105-16, 105-16R.
174 Logan to Archibald, June 10, 1947 LAC, RG 17, 3296, file 559-17.
175 Archibald to Deputy Minister, December 8, 1945, LAC, RG 17, 3296, file 559-17.
176 Ibid., May 17, 1946, LAC, RG 17, 3296, file 559-17.
177 Ibid., October 10, 1947, LAC, RG 17, 3296, file 559-17.
178 J.S. Parker, "Maritime Marshland Reclamation," June 21, 1956, NSDA.
179 *Maritime Marshland Rehabilitation Act*, SC 1948, c. 61.
180 Canada, Parliament, House of Commons, *Debates*, 20th Parl., 4th Sess., 5 (1947–48): 4503. There was a single MMRA project in Prince Edward Island, so it was effectively a program pertinent to New Brunswick and Nova Scotia.
181 Ibid., 4509.
182 Ibid., 4511.
183 *Marshland Reclamation Act*, SNS 1949, c. 2; *Marshland Reclamation Act*, SNB 1949, c. 22.
184 *Marsh Act*, SNS 1900, c. 12, s. 16. "Manurance" is an obsolete term for "cultivation."
185 *Marshland Reclamation Act*, SNS 1949, c. 2, ss. 3–4. Marsh bodies that chose not to take part in the new system would continue to be ruled by the old legislation, which continued to countenance such practices as tiding.
186 Archibald to Deputy Minister, October 10, 1947, LAC, RG 17, 3296, file 559-17.
187 Assessment Roll, January 20, 1950, Acadia University Archives, Grand Pre Marsh Body Fonds, 2007.032-GMB/6, https://chipmanscorner.acadiau.ca/islandora/object/research%3A2058.
188 New Brunswick, *Synoptic Report of the Proceedings of the Legislative Assembly of New Brunswick*, April 7, 1949, 164.
189 Dentiballis Marsh Body, Minutes, October 10, 1949.
190 Minutes of Grand Pre Common Field, March 26, 1949, to March 25, 1950, King's County Museum (Kentville, NS), MS11. In the case of the Falmouth Great Dyke Marsh Body, a folded-over page was the only indication in the minute book that began in 1907 that a change had taken place.
191 Grand Pre Marsh Body, Executive Committee, Minutes, November 8, 1949, Acadia University Archives, Grand Pre Marsh Body Fonds, 2007.032-GMB/2, https://chipmanscorner.acadiau.ca/islandora/object/research%3A1950.
192 Curry to Parker, November 15, 1949, December 19, 1949, Acadia University Archives, MMRA Fonds, 1998.001/1/2.
193 Parker to Curry, December 21, 1949, Acadia University Archives, MMRA Fonds, 1998.001/1/2.

Chapter 2: Reconstruction

1. Minister of Agriculture to Cabinet, June 24, 1949, LAC, RG 17, vol. 3443, file 4138-11; Taggart to Gardiner, May 23, 1949, LAC, RG 124, series A-3, vol. 25, file 165-M5; MMRA Advisory Committee meeting (hereafter MMRA AC), May 9, 1949.
2. MMRA AC, June 14, 1950; August 29, 1952.
3. Ibid., September 23–25, 1952.
4. Ibid.
5. Minutes of Special Meeting of the Marshland Utilization Committee, November 8, 1949, LAC, RG 124, series A-3, vol. 25, file 165-M5.
6. *Amherst Daily News,* February 8, 1950.
7. MMRA AC, October 21, 1949.
8. Maritime Marshland Rehabilitation Administration (MMRA), *Annual Report, 1949–50,* 5; R. Peterson, "Notes on Visit to Maritime Marshland Rehabilitation Projects, August 15th–19th, 1951," NBDA.
9. In 1961, Parker reported to a Senate committee that the 80,000 acres of protected dykeland formed part of farms that contained 450,000 acres. These figures also took into account the tidal dams that had been constructed by then, but if the same ratio were applied to just the land protected by the dykes and aboiteaux, those farms would contain roughly 280,000 acres. Canada, Parliament, Senate, Standing Special Committee on Land Use, April 26, 1961.
10. MMRA AC, September 19-21, 1960.
11. MMRA *Annual Reports* include data regarding marshland that had been incorporated into marsh bodies, but where no protective structures existed. In 1959, the last year for which information at the marsh body level is available regarding unprotected marsh, over 2,000 acres along the Cumberland Basin were "out to sea." The creation of National Wildlife Areas on land that had at one time been drained is discussed at length in Chapter 4.
12. Scott, *Seeing Like a State,* 315–16.
13. Reilly, "The General Strike in Amherst," 58.
14. Ibid., 59–60.
15. *Family Herald and Weekly Star,* March 22, 1950. A similar text was published in the *Amherst Daily News,* February 3, 1950.
16. *Family Herald and Weekly Star,* March 22, 1950.
17. *Amherst Daily News,* February 3, 1950.
18. *Family Herald and Weekly Star,* March 22, 1950.
19. *Instructions for Aboiteau Construction,* NSDA, Windsor.
20. When the provinces took over responsibility for the dykes, aboiteaux, and dams in 1970, the MMRA generally transferred from Amherst only documents pertinent to the upkeep of those structures. Information and images concerning what went on at Amherst rarely survived, as far as I can tell, because they were not useful for carrying out work on the ground. Another glimpse of the interior of the MMRA facilities can be found in *Les aboiteaux,* as Placide is given a tour of the Amherst operation.
21. *Amherst Daily News,* February 3, 1950.
22. L.W. McCarthy, Progress Report on the Activities of the Soil Mechanics Division of the Engineering Department, for the Period October 1, 1949 – March 31, 1952, NSDA.
23. MMRA AC, "MMRA Tentative Drainage Standards," Appendix to Meeting, June 7–8, 1951.

24 Interviews with Ernie Partridge, Dorchester, NB, September 2015, May 2016. Ernie was the only former MMRA employee still alive and able to describe the operations at Amherst.
25 Taggart to Gardiner, August 18, 1948, LAC, RG 17, 3296, file 559-17.
26 Of all the professionals employed by the MMRA, Frail's career is the best documented, thanks to a collection housed at the Acadia University Archives (2015.035-FRA), which includes documents, images, and video. Further video and images are contained in the George Frail collection held by the Annapolis Heritage Society, at the O'Dell House Museum, Annapolis Royal, NS.
27 Interview with John Waugh, Amherst, NS, May 4, 2016.
28 Harry Thurston, Constant Desplanque Obituary, *Globe and Mail*, July 13, 2011; Constant Desplanque, "The Story of My Life" (unpublished manuscript), 20. I received a copy of this manuscript from David Mossman, a professor at Mount Allison University, who published scientific articles with Desplanque after the latter retired from the MMRA in 1982. See, for instance, Desplanque and Mossman, "Tides and Their Seminal Impact on the Geology, Geography, History, and Socio-Economics of the Bay of Fundy, Eastern Canada."
29 Regarding Byers, Nova Scotia Agricultural College, *Calendar 1942–1943*, https://dalspace.library.dal.ca/bitstream/handle/10222/43747/nsac_calendar_1942-1943.pdf?sequence=1.
30 Interview with R.H. Palmeter, Acadia University Archives, MMRA Fonds, 1998.002/1/4. We met Byers in Chapter 1 as leading the Nova Scotia Marsh Survey. The construction of this dyke through the use of machinery and the problems connected with that project are discussed in Chapter 1.
31 *Amherst Daily News*, February 3, 1950. Palmeter's resignation as a commissioner for the Grand Pre Marsh Body is recorded in Minutes of Grand Pre Common Field, July 20, 1949, King's County Museum (Kentville, NS), MS11.
32 *Instructions for Aboiteau Construction*, NSDA, Windsor. Fred Palmeter was an experienced hand at dyke construction, having served as one of the models for the character of Placide in the film *Les aboiteaux*.
33 MMRA report on Grand Pre Marsh Body, September 10, 1949, Acadia University Archives, MMRA Fonds, 1998.001/1/2.
34 *Instructions for Aboiteau Construction*, NSDA, Windsor.
35 Loo, "High Modernism," 45–46.
36 Fitzgerald, *Every Farm a Factory*, 22.
37 Ibid., 78–79.
38 Parker, McCarthy, and McIntyre, "Reclamation of Tidal Marshlands"; McIntyre and Desplanque, "Silt Deposition and Erosion"; McIntyre and Palmeter, "Protection of Earth Structures by Vegetative Cover," 14–17.
39 Outline for MMRA Seminar Series, 1958, NSDA.
40 Ibid.
41 Ibid.
42 Nova Scotia Marshland Reclamation Commission, minutes of meeting, September 26, 1949, NSDA. In a sense, *Les aboiteaux* was that film.
43 C.A. Banks, "Photography in Surveying," May 8, 1957, NSDA. Banks emerged as "Chief of Technical Services," at Amherst in the late 1960s, but his precise title in 1957 is not known.
44 Lévesque to Maurice Sauvé, May 12, 1965, LAC, MG 32, vol. 13, file 12.

45 In still other cases, documented in the second part of this book, the experts became so convinced that their actions could cause no harm that they ignored warnings about constructing dams that obstructed a number of the region's major rivers.
46 General Procedures, Maritime Marshland Reclamation, NSDA, Windsor.
47 Ibid.
48 Taggart to Gardiner, May 23, 1949, LAC, RG 124, vol. 25, file 165-M5.
49 In New Brunswick, 35 marsh bodies had been incorporated by 1950, a number that rose to 47 three years later, by which time the province stopped publishing information about new bodies being brought into the fold. In Nova Scotia, more than 50 bodies signed up in the first year, or roughly half of the 115 that would be incorporated by 1958, at which point the province's Department of Agriculture could announce that there had been "a definite slackening-off in the programme." Nova Scotia, Department of Agriculture, *Annual Report for Year Ending 31 March 1958*, 12. The data on incorporation of marshland across the region come from MMRA, *Annual Report for Year Ending 31 October 1964*, 25.
50 Interview with Jim Bremner, Falmouth, NS, July 13, 2017.
51 J. Adrien Lévesque to Harry Hays, May 12, 1965, LAC, MG 32, vol. 13.
52 A.E. Black to Parker, October 26, 1949, NBDA.
53 C.L. Beasley to C.E. Henry, July 20, 1950, NSDA.
54 When I came across the provincial marshland archives, largely untouched for forty years, they were organized according to this numerical system.
55 Scott, *Seeing Like a State*, 11, 346.
56 Parker to Leslie Harvey, March 30, 1950, Acadia University Archives, MMRA Fonds, 1998.001/1/2.
57 Agreement between Grand Pre Marsh Body and Nova Scotia Minister of Agriculture, October 5, 1949, NSDA. This was a standard contract used for all agreements between the province and individual marsh bodies.
58 For instance, in the case of Nova Scotia's Fort Lawrence Marsh Body, work that was authorized in 1958 called for an investment of $82,000 for protective structures and $135,000 for the ditches. MMRA AC, March 22, 1957.
59 By 1961, the MMRA had eliminated 694 aboiteaux through "redesigning drainage systems." This subject is discussed at greater length in Chapter 3. Canada, Parliament, Senate, Standing Special Committee on Land Use, testimony of J.S. Parker, April 26, 1961.
60 MMRA AC, "Report of the Committee Appointed to Bring Forward Recommendations Concerning Drainage Problems," February 19, 1952. This matter festered throughout the 1950s and resulted in appointment of a Marshland Drainage Committee in 1959, whose work is reported in Canada, Department of Agriculture, *Machines for Marshland Ditching*.
61 General Procedures, Maritime Marshland Reclamation, NSDA, Windsor.
62 See, for instance, the "Plan of Proposed Dyke on Grand Pre Marsh," which has the names of proprietors inscribed within the borders of the properties. Acadia University Archives, Grand Pre Marsh Body Fonds, 2007.032-GMB/44, https://chipmanscorner.acadiau.ca/islandora/object/research%3A1952?search=dyke%2520grand%2520pre. Maps of both the Bear Island and West Body Marshes in New Brunswick provide numerous references to unnumbered "aboideaux." Mount Allison University Archives, Donald Harper Fonds, 7740/2/6/4, 7740/2/6/7.
63 Scott, *Seeing Like a State*, 44–47. State planners who decided to remove the residents of seven New Brunswick villages to create Kouchibouguac National Park similarly produced

Notes to pages 92–99　　245

maps that facilitated the expropriation process, inscribing a unique number on each parcel of land (Rudin, *Kouchibouguac*, 89–93). These expropriation maps figure prominently in the website "Returning the Voices to Kouchibouguac National Park," http://returningthevoices.ca.
64　Parker to G.J. Matte, September 21, 1949, LAC, RG 124, series A-3, vol. 25, file 165-M5.
65　Ibid. For an overview and critique of this stereotype of Atlantic Canadians, see MacKinnon, "Five Simple Rules for Saving the Maritimes."
66　Parker to Matte, September 21, 1949, LAC, RG 124, series A-3, vol. 25, file 165-M5.
67　Report of the Marshland Utilization Meeting, Amherst, February 7–8, 1950, NSDA.
68　MMRA AC Meeting 29, March 27–28, 1958; Appendix: Information Prepared for the Maritime Marshland Reclamation Committee, J.A. Roberts, "Utilization of New Brunswick Marshlands," September 24, 1957. The decline in the value of land was discussed in Chapter 1.
69　Report of the Marshland Utilization Meeting, Amherst, February 7–8, 1950, NSDA.
70　MMRA AC, March 17–18, 1954.
71　Fitzgerald, *Every Farm a Factory*, 77–78.
72　Haase and Packman, *Marshland Utilization;* Haase, *Some Economic Aspects*. As we saw in Chapter 1, these studies based on surveys of dykeland farming are invaluable since it is impossible to distinguish dykeland from other agricultural land in the aggregated data provided by the Census of Canada.
73　MMRA AC, February 22–23, 1951.
74　Ibid.
75　Ibid.
76　Ibid.
77　A.L. Stevenson, Memo on Report of Gordon Haase on Marshland Reclamation, April 17, 1951, LAC, RG 17, vol. 3443, file 4138-23.
78　J.G. Taggart to R.B. Bryce, July 10, 1950, LAC, RG 17, vol. 3443, file 4138-15.
79　Meeting re Maritime Marshland Rehabilitation, January 18, 1954, NSDA.
80　MMRA AC, February 18, 1954.
81　Ibid., March 17–18, 1954.
82　*Cost-Benefit Analysis, Nova Scotia and New Brunswick Marsh Projects,* MMRA, June 28, 1954, NSDA.
83　Canada, Parliament, Senate, Standing Special Committee on Land Use, April 26, 1961.
84　Appended to the minutes of the Advisory Committee were forms providing information about projects that were accepted. However, those forms were not appended for files that were rejected.
85　MMRA AC, March 17–18, 1954.
86　Ibid., July 13, 1954, C.E. Henry, "Report on Centre Burlington Marsh Body." The cost-benefit figure is from *Cost-Benefit Analysis, Nova Scotia and New Brunswick Marsh Projects,* MMRA, June 28, 1954, NSDA.
87　MMRA AC, July 13, 1954, Appendix D: F.W. Beattie, "In the Matter of the Marshland Reclamation Act."
88　Ibid.
89　Parker to F.W. Walsh, Nova Scotia Deputy Minister of Agriculture, September 24, 1955, LAC, RG 124, vol. 26, file 165-M5, pt. 3.
90　MMRA AC, February 22–23, 1951.

91 MMRA, "Aboiteau Construction," NSDA, Windsor.
92 *Amherst Daily News,* February 3, 1950. The ability of the agency to dispatch such machinery to a job site was dramatically presented in the film *Les aboiteaux,* discussed at length in the Prologue.
93 Parker to Taggart, May 23, 1950, LAC, RG 17, vol. 3443, file 4138-23; Parker, "Maritime Marshland Reclamation," June 21, 1956, NSDA.
94 MMRA Recommendation for work on NS 8, Grand Pre Marsh Body, December 19, 1949, NSDA.
95 Minutes of Executive Committee, February 22, 1950, Acadia University Archives, Grand Pre March Body Fonds, 2007.032-GMB/2, https://chipmanscorner.acadiau.ca/islandora/object/research%3A1950.
96 Parker to Leslie Harvey (secretary of Grand Pre Marsh Body), March 2, 1950, Acadia University Archives, MMRA Fonds, 1998.001/1/2.
97 These efforts to close off the region's rivers are the focus of the second part of this book.
98 The Preface provides a complete list of the marsh body minutes I was able to consult, all in Nova Scotia. The situation in New Brunswick is discussed below.
99 Falmouth Great Dyke Marsh Body, Minutes, January 6, 1947; February 2, 1953 (private collection, Philip Davison).
100 Subsequently, there were several meetings in the 1970s, the last of which was in 1979. The marsh body effectively had nothing to do following completion of the Avon River Causeway in 1970, a subject discussed at length in Chapters 3 and 4.
101 The minutes for the Dentiballis Marsh Body (near Annapolis Royal) began in 1887, while those that I found for the Victoria Diamond Jubilee Marsh Body (near Truro) stretch back to 1898. The Bishop Beckwith Marsh Body was founded at the same time as the one at nearby Grand-Pré, immediately after removal of the Acadians.
102 For an introduction to this initiative, see Young, "The Programme of Equal Opportunity."
103 The situation prior to the 1960s is described in Lewey, Richard, and Turner, *New Brunswick before the Equal Opportunity Program.*
104 SNB 1966, c. 77.
105 *Synoptic Report of the Proceedings of the Legislative Assembly of New Brunswick,* June 2, 1966.
106 Ibid.
107 Roberts to R.D. Gilbert, February 3, 1967, Provincial Archives of New Brunswick (PANB), RS 123, Agricultural Engineering Branch, file 2b.
108 MMRA, Report on "Future Maintenance of Protective Works," 1952, NSDA. Given the MMRA's preoccupation with reducing the number of aboiteaux, its willingness to replace aboiteaux that were still in relatively good condition spoke volumes about the agency's negative view of the marsh bodies.
109 Ibid.
110 MMRA, Circular letter, No. 35, June 12, 1953, NSDA. By saying that it wanted to avoid being "dictatorial," the MMRA was effectively recognizing that its actions fit the bill.
111 Memorandum from Parker to District Engineers, July 7, 1953, NBDA. This circular was also discussed by the Nova Scotia Marshland Reclamation Commission (NSMRC), which found that the document was "very liberal insofar as the Marsh Bodies were concerned." It is not clear what they meant by "liberal." NSMRC, Minutes of Meeting, June 15, 1953, NSDA.
112 *Maritime Marshland Rehabilitation Act,* SC 1948, c. 61.

113 Agreement between Grand Pre Marsh Body and Nova Scotia Minister of Agriculture, October 5, 1949, NSDA.
114 "Summary Minutes of Meeting on Maintenance of Maritime Marshland Protective Works," September 1–2, 1959, NSDA.
115 Ibid. There had been a meeting of the provincial leaders on the dykeland dossier in 1956, in anticipation of the federal government's desire to transfer responsibility. In the report of the meeting, the first item read: "It was generally agreed that it would not be wise to place responsibility for the dykes and aboiteaux entirely in the hands of the marsh owners." Report of Meeting to Consider Maintenance of Marshland Projects, September 11, 1956, PANB, RS 124 F1, m2.
116 Ibid.
117 MMRA, *Marshland Utilization Statistics for the Year 1964*. On the exceptional situation in the Kentville-Windsor area, see Haase and Packman, *Marshland Utilization*, 13–18. The 1964 study divided marsh into two categories, one that was suitable for agriculture and one that was not. When added together, the total area of marsh is roughly the same as the area in the MMRA annual reports for "protected marsh."
118 MMRA, *Marshland Utilization Statistics for the Year 1964*.
119 Lévesque to Harry Hays (federal Minister of Agriculture), May 12, 1965, LAC, Maurice Sauvé Fonds, MG 32 B 4, vol. 13, file 12.
120 Ibid.
121 G.C. Retson, "Marshland Farming," 20–26.
122 Parson, "Regional Trends of Agricultural Restructuring," 346.
123 "Summary Minutes of Meeting on Maintenance of Maritime Marshland Protective Works," September 1–2, 1959, NSDA.
124 J.A. Roberts, "Marshland Maintenance and Upgrading Program," n.d., NBDA.
125 "Agreement for the Reclamation of Marshlands in New Brunswick under the Maritime Marshland Reclamation Act," March 31, 1966, PANB, RS 123, Agricultural Engineering Branch, Marshlands Reports. A parallel agreement was signed between Ottawa and Halifax. The "official" history of dyke management noted that in 1966, "Ottawa startled some Maritime officials by announcing its intention to transfer the responsibility for dykeland protection to the provinces by 1970." But this could hardly have startled the provincial officials who had been involved in discussions about the transfer since the late 1950s. Nova Scotia, Department of Agriculture and Marketing, *Maritime Dykelands*, 89.
126 *Ottawa Journal*, June 28, 1961. Parker became the director general of the department's Administration Branch in Ottawa. He was succeeded at the MMRA by Gideon Matte, who knew the file well from his earlier connection with the PFRA.
127 Canada, Parliament, Senate, Standing Special Committee on Land Use, April 26, 1961.
128 *Canadian Business*, 34 (July 1961): 14.
129 Canada, Parliament, Senate, Standing Special Committee on Land Use, April 26, 1961.
130 Buckley and Tihanyi, *Canadian Policies for Rural Adjustment*, 12–13.
131 Ibid., 87–88.
132 Ibid., 91. Their conclusion about the benefits of the dams could not take into account the significant environmental costs that emerged, particularly in terms of the Petitcodiac Causeway, which had not yet been completed at the time of their study. We will return to those costs in Chapter 4.
133 Ibid., 92.

134 Raymond Scovil (Assistant Deputy Minister of Agriculture) to Robert Moreland (Treasury Board), October 9, 1969, PANB, RS 123, Agricultural Engineering Branch, Correspondence; Roberts to Scovil, September 29, 1969, PANB, RS 124, G39a.
135 Roberts's counterpart in Nova Scotia had comparable credentials. In addition to heading the province's Marshland Reclamation Commission, C.E. Henry was a professor of agricultural engineering at the Nova Scotia Agricultural College and served as regional director for the Canadian Society of Agricultural Engineers.
136 With regard to the creation of Kouchibouguac National Park, there was a similar divide between federal officials who viewed the people being removed from their lands as abstractions and those from New Brunswick, who sought to soften the impact of removal. See Rudin, *Kouchibouguac*, 66.
137 Roberts to Scovil, September 29, 1969, PANB, RS 124, G39a; Roberts to R.D. Gilbert (Deputy Minister of Agriculture), February 13, 1968, PANB, RS 123, Agricultural Engineering Branch, Correspondence.
138 Rural Development Branch, Department of Forestry and Rural Development, "Report on the MMRA program Arising out of Observations Made in Special Study No 7 by the Economic Council of Canada entitled Canadian Policies for Rural Adjustment by Buckley and Tihanyi, 1967," February 1968, PANB, RS 124, G39a.
139 Interview with Ernie Partridge, Dorchester, NB, September 2, 2015.
140 Organization Division, Advisory Services Branch, Civil Service Commission, "Organization Study of ARDA and MMRA in the Department of Forestry," December 1964, LAC, Maurice Sauvé Fonds, MG 32 B 4, vol. 2, file 11.
141 Buckley and Tihanyi, *Canadian Policies for Rural Adjustment*, 93. In 1965, ARDA was renamed (with the same acronym) the Agricultural and Rural Development Administration.
142 MMRA, *Annual Report, 1963–64*, 8. As part of its move away from dykeland protection, the MMRA was transferred out of the Department of Agriculture in February 1964. It was moved to the Department of Forestry, where ARDA was housed at the time. In June 1964, as part of its new, larger mandate, the MMRA was asked to "make a study of the Gold River [in Nova Scotia] to determine the feasibility of increasing the Atlantic Salmon potential of the river" (MMRA, "Gold River Study," June 1964, LAC, RG 124, accession 1984–85/617, box 6, file 1050-1). This same river had achieved a certain notoriety in the early 1960s when Nova Scotia Power and Light proposed to dam it. Protests led to a government study that resulted in the project's shelving. Mark Leeming describes this as one of the first instances in Nova Scotia of successful mobilization by environmentalists against ill-conceived projects. Leeming, *In Defence of Home Places*, 13–16.
143 Submission to Maurice Sauvé from Amherst Town Council, Amherst Board of Trade, and Amherst Industrial Commission, September 9, 1966, LAC, Maurice Sauvé Fonds, MG 32, vol. 13, file 7.

CHAPTER 3: DAM PROJECTS

1 Laurie Anderson's Meeting, November 12, 1952, NBDA. Anderson was part of a large family that was prominent in the marsh-owning community around Sackville.
2 Worster, *Rivers of Empire*, 264.
3 Archibald to John A. McDonald (Nova Scotia Minister of Agriculture), October 14, 1943, LAC, RG 17, file 3295, pt. 1.

4 MMRA, "Future Maintenance of Protective Works," 1953, NSDA.
5 Scrapbook of NS 15, Isgonish River, Acadia University Archives, George Frail Fonds, 2015.035-FRA/18.
6 In discussing the construction of "a dam and aboiteau," the MMRA Advisory Committee reflected its own confusion about the difference between the two types of structures (MMRA AC, March 21, 1952).
7 To add to the terminological confusion, some MMRA projects referred to as dams were really aboiteaux. For instance, Nova Scotia's Nappan "Dam," completed in 1959, was really a large aboiteau, which (much like George Frail's Isgonish aboiteau discussed above) allowed water to drain out instead of being retained in a headpond. I owe a large debt to Gary Gilbert, the aboiteau superintendent in the Amherst region, with an office a stone's throw from the Nappan structure, who helped clarify my understanding of how such structures worked.
8 Banks, *Preliminary Report on Tidal River Siltation*, 8.
9 Laurie Anderson's Meeting, November 12, 1952, NBDA. As we will see with regard to some of the later dams, locating a dam upstream was ill-advised because it would result in the buildup of silt that had washed up the river before settling out. While the farmers were not trained in the dynamics of dam construction, the MMRA engineers were. Nevertheless, they made no comment in this regard in rejecting the idea of locating the Tantramar Dam upstream, a practice they implemented in constructing both the Petitcodiac and Avon River dams far from their mouths.
10 Ibid.
11 MMRA, "Summary Report on Project N.B. 22, Tantramar River Survey," February 15, 1953, 45–46, NBDA; MMRA, "Summary Report on Project N.S. 7, Annapolis River Survey," January 30, 1954, 25–26, NSDA.
12 Scott, *Seeing Like a State*, 316.
13 Loo, "High Modernism," 46.
14 Ibid., 49.
15 MMRA, *Annual Reports*.
16 Minutes of Meeting ... Convened to Consider Problems Relating to Marshland Rehabilitation, September 1943, LAC, RG 17, file 3295, pt. 1.
17 MMRA AC, July 22, 1949; Parker to G.J. Matte, November 17, 1949, NBDA.
18 New Brunswick, Department of Agriculture, "Rehabilitation of the Shepody Area, Albert County, N.B," 1951, NBDA.
19 J.A. Roberts, "Report on Urgent Dyke repairs needed in Albert Co, NB," LAC, RG 19, vol. 432, file pt. 1, file 105-16.
20 New Brunswick, Department of Agriculture, "Rehabilitation of the Shepody Area, Albert County, N.B," 1951, NBDA.
21 J.S. Parker, Memorandum on General and Engineering Aspects of the Shepody River Project, February 21, 1951, NBDA.
22 As of the end of March 1951, the MMRA's total expenditures came to $1,339,705. MMRA, *Annual Reports*, 1949–50, 1950–51.
23 *Moncton Transcript*, March 10, 1951; Report on Meeting of Marshowners in the Shepody Area, March 7, 1951, NBDA.
24 New Brunswick Marshland Utilization Committee, "Utilization of Marshlands in the Shepody Area, Albert Co. NB," 1951, NBDA.

25 Ibid.
26 In a similar manner, when the Canadian and New Brunswick governments worked together to create Kouchibouguac National Park, they transformed the landscape at the same time that they sought to "rehabilitate" the population. Rudin, *Kouchibouguac*, 106–13.
27 New Brunswick, Department of Agriculture, "Rehabilitation of the Shepody Area, Albert Co., N.B," 1951, NBDA.
28 MMRA AC, June 7–8, 1951.
29 Report on Meeting of Marshowners in the Shepody Area, March 7, 1951, NBDA.
30 MMRA AC, June 7–8, 1951.
31 *Moncton Daily Times*, June 5, 1954. A similarly heroic article was published in the same newspaper on August 23, 1954.
32 MMRA AC, April 6, 1956.
33 Ibid.
34 Banks, *Preliminary Report on Tidal River Siltation*, 4.
35 Ibid., 11.
36 Notes from the July 15, 1958 Meeting with the MMRA to discuss fish facilities in the Annapolis Aboideau, NSDA.
37 MMRA AC, February 22–23, 1951.
38 Ganong, "The Vegetation of the Bay of Fundy Salt and Diked Marshes," 180; Thompson's exploits and his memory in the region are also discussed in Scobie, *People of the Tantramar*, 25–26; and Cunningham, "Thompson, Toler (Tolar)."
39 Archibald's anti-tiding views were discussed in Chapter 1.
40 Roberts to Parker, December 12, 1950, NBDA (strikethrough in original).
41 Scott, *Seeing Like a State*, 254.
42 Parker to Taggart, January 24, 1950, NBDA.
43 Treasury Board, February 18, 1950, LAC, RG 17, vol. 3443, 4138-15-5; MMRA AC, January 29–30, 1953.
44 *Summary Report on Project NB 22*, 7.
45 MMRA AC, January 28–29, 1953.
46 New Brunswick Marshland Reclamation Commission, Meeting, March 25, 1953, NBDA; MMRA AC, March 27, 1953.
47 Report on Meeting of Marshowners in the Tantramar Area, April 29, 1953, NBDA.
48 Ibid.
49 "Marshowners Discuss Protection of Tantramar Area," April 1953, PANB, RS 124 F1 m2.
50 *Sackville Tribune-Post*, 12 June 1953.
51 New Brunswick Marshland Reclamation Commission, Meeting, August 12, 1953, NBDA.
52 King to Roberts, July 29, 1953, NBDA.
53 Roberts to Anderson, August 4, 1953, NBDA.
54 New Brunswick Marshland Reclamation Commission, Meeting, August 12, 1953, NBDA.
55 Fawcett to Parker, August 28, 1953, NBDA.
56 *Sackville Tribune-Post*, August 21, 1953.
57 R. Ernest Estabrooks, letter to editor, *Sackville Tribune-Post*, August 27, 1953.
58 New Brunswick Marshland Reclamation Commission, Roberts to C.B. Sherwood (New Brunswick Minister of Agriculture), December 21, 1953, NBDA.
59 Log Lake Tract Brief on Tiding, presented to C.B. Sherwood, New Brunswick Minister of Agriculture, April 1954, NBDA.

60 Tiding in Log Lake Marsh, N.B. 29, NBDA. The term "clapper" is an Anglicized version of the Acadian term for the gate, *le clapet*.
61 Ibid.
62 Parker to Taggart, May 1, 1954; Parker to G.J. Matte, May 4, 1954, NBDA.
63 For the history of the Trans-Canada Highway, see Francis, *A Road for Canada*.
64 The correspondence concerning this arrangement can be found in PANB, RS 693, Tantramar Dam File. New Brunswick premier H.J. Flemming proposed the arrangement to federal agriculture minister J.L. Gardiner, February 20, 1957; Gardiner's acceptance of the arrangement is in his response, March 6, 1957. Ottawa agreed to pick up half of the province's costs in a letter from the minister of public works, Robert Winters, to Flemming, March 21, 1957. As Winters put it: "Our contribution [to the highway over the Tantramar Dam] would be under the general principle of the Trans-Canada Highway Agreement by which we would pay 50% of the cost to the province."
65 MMRA AC, July 22, 1949. The prospects for combining aboiteau and bridge reconstruction surfaced regularly in Advisory Committee minutes. For instance, at a meeting in 1952, the committee discussed building structures across both the Isgonish and Gaspereau Rivers in Nova Scotia. With regard to the latter project, Parker explained how "one large aboiteau, also serving as a highway bridge, could protect marsh along a river." MMRA AC, March 21, 1952.
66 Marsh Owners' Association of Annapolis County to Louis St-Laurent et al., January 13, 1946, LAC, RG 2 1990-91/154 105, file D-10-2A. The same file includes further expressions of support in the 1950s: Marsh Owners' Association of Annapolis County to federal Minister of Agriculture, October 6, 1955; and resolution of Annapolis Royal Town Council, November 10, 1955.
67 B. Russell, "Marsh Lands of the Maritime Provinces, Report of Reconnaissance," November 22, 1943, LAC, RG 17, 3295, pt. 1. As for the reference to tidal power, a generating station opened in 1984, placing it far beyond the scope of this book.
68 Treasury Board, September 14, 1949, LAC, RG 17, vol. 3443, file 4138-15-2. In authorizing the Tantramar feasibility study, the Treasury Board asked only whether "it is feasible to construct a number of dams and aboiteaux and enable the reclamation of additional lands for agricultural purposes." Treasury Board, February 18, 1950, LAC, RG 17, vol. 3443, file 4138-15-5.
69 The Shepody reservoir has a capacity of 580 million cubic feet, while the Annapolis reservoir could hold nearly 3 billion cubic feet.
70 Parker to A.W. Mackenzie, March 20, 1953, NSDA.
71 Ibid.
72 According to the MMRA's chief engineer, the construction of the dam played a role in the collapse of the bridge. J.D. Conlon noted in a letter to Parker: "I believe the heavy turbulent flows, resulting from the Annapolis Dam construction operations, were responsible for the collapse of the bridge last June [1960]." While Conlon recognized that the "bridge foundations were known to be weak," he also noted that "these flows were substantially more than 'the straw that broke the camel's back.'" Conlon to Parker, April 7, 1961, NSDA.
73 Meeting re Maritime Marshland Rehabilitation, January 18, 1954, NSDA.
74 MMRA AC, October 20, 1954.
75 Parker to Taggart, February 2, 1955, NSDA.
76 Ibid.

77 By agreement, Ottawa and Halifax were to divide costs at a ratio of 1.75:1. See PC 1957-5/626, May 3, 1957.
78 Canada, Department of Agriculture, Information Division, News release K-29, June 29, 1960, NSDA.
79 MMRA, "Summary Report of Project N.S. 7, Annapolis River Survey," January 30, 1954, NSDA.
80 Halifax *Chronicle Herald,* April 7, 1952.
81 Parker to J.C. Belliveau (Nova Scotia Deputy Minister of Highways and Public Works), March 27, 1952. I received this document and the one in the following note from Charles Curry, a member of the Horton Marsh Body (NS 72). I would not even have known about this aborted MMRA project if it had not been for him, so I am very grateful.
82 Parker to W.H. Pipe (MLA), April 23, 1953.
83 In 1955, again feeling pressure from those who felt strongly that "the construction of an aboiteau [across the Gaspereau] would seriously interfere with the fishing on the river," the Advisory Committee approved an investment of $68,000 for reconstruction work. MMRA AC, December 16, 1955.
84 R.L. Butler, Abstract from weekly reports, Albert, N.B., October 11, 1958, NSDA, Nappan.
85 B.W. Hamer, Memo regarding Aboideau Dams, 1958, NSDA, Nappan.
86 "Notes from the July 15 1958 meeting with the Maritime Marshland Rehabilitation Administration at Amherst N.S. to discuss fish facilities in the Annapolis Aboiteau," NSDA. The hydraulics lab mentioned here was the same one described in Chapter 2 where Ernie Partridge worked during the winter.
87 R.W. Logie to J.S. Robertson (Nova Scotia Salmon Anglers Association), March 26, 1965, LAC, RG 23, file 5903-82 2-1.
88 Ibid.
89 Alwyn Cameron to Parker, October 18, 1950, NBDA.
90 MMRA, "Summary Report of Project N.S. 7, Annapolis River Survey," January 30, 1954, NSDA.
91 Canada, Public Health Engineering Division, *Aspects of Pollution,* 18.
92 Brief of the Town of Bridgetown Dealing on the Sewage Problem Created and Now Existing as a Result of the Annapolis River Causeway, August 5, 1960, NSDA.
93 Parker to J.L. Wickwire (Nova Scotia Deputy Minister of Highways), July 22, 1960; Parker to S.C. Barry (Nova Scotia Deputy Minister of Agriculture), October 14, 1960; Parker to Wickwire, November 15, 1960, NSDA.
94 Canada, Public Health Engineering Division, *Aspects of Pollution,* 18.
95 C.E. Daniels and Son to MMRA, August 5, 1957, NSDA.
96 H.H. Smofsky to Parker, April 1, 1960, NSDA.
97 MMRA AC, June 9, 1965.
98 Daye, "Booming, Bustling Moncton," 15–21.
99 Spence-Sales, *Moncton Renewed,* 1: n.p.
100 Spence-Sales, *Guide to Urban Dispersal,* n.p.
101 These were the towns classified by the census as forming part of the Moncton urban area in 1961.
102 Spence-Sales, *Guide to Urban Dispersal,* n.p.
103 Scott, *Seeing Like a State,* 347.
104 Moncton City Council, Meeting of Committee of the Whole, January 5, 1960, City of Moncton Archives.

Notes to pages 150–56

105 O.J. McCulloch and Co. to Robert Winters, May 9, 1956, LAC, RG 19-F-2, vol. 4427, file 9320-02, pt. 1.
106 Parker to C.S. Williams (Chief Engineer, Department of Public Works), September 5, 1956, NBDA. With regard to the proposal here to block the mouth of the Memramcook River, it resurfaced in the late 1960s with a plan to obstruct it with a causeway well upstream, much as the Petitcodiac Causeway obstructed that river far from its mouth. At the final meeting of the Advisory Committee, a feasibility study was tabled that outlined a variety of benefits from the Memramcook project, of which marshland protection was relatively insignificant. In this as in other contexts, by the late 1960s the MMRA was comfortable with supporting projects beyond the protection of dykeland (MMRA AC, October 8, 1969). The Memramcook dam was completed after the MMRA had closed down, and so is beyond the scope of this book. The river – now diminished because of the causeway – can be seen behind Ernie Partridge's property in Dorchester, downstream from the structure. As I stood with him during an interview, he looked out at the river, where he had first learned about the power of tides, and shook his head in frustration over what had been done to the Memramcook.
107 Harvey, "Death Watch on the Petitcodiac," 38–39.
108 Moncton City Council, Meeting of Committee of the Whole, January 5, 1960, City of Moncton Archives.
109 Tricker, *Bores, Breakers, Waves and Wakes*, 33–34.
110 Dalton, "Fundy's Prodigious Tides," 229.
111 Ibid., 230.
112 On the back of this postcard, which I acquired via eBay, the tourist wrote to someone in Hartford, Connecticut, indicating that she was on a tour that included both the tidal bore and a visit to Hopewell Rocks, still a tourist attraction in the early twenty-first century, forty kilometres downstream from Moncton on the Petitcodiac. An image search of this postcard returns numerous other postcards depicting the tidal bore.
113 CBC New Brunswick, "U.S. Surfers Recall Adventure of Riding Moncton's 'Tiny' Tidal Bore in 1967," August 24, 2017, https://www.cbc.ca/news/canada/new-brunswick/american-surfers-petitcodiac-river-tidal-bore-moncton-1.4260188. Surfing has now returned to the Petitcodiac, following the opening of the causeway's gates in 2010. The twenty-first century surfers are discussed in Chapter 4.
114 Lynch, "Tidal Bores," 156.
115 Moncton City Council, Meeting of Committee of the Whole, January 5, 1960, City of Moncton Archives. Parker's use of the term "naturally" was unintentionally ironic.
116 Ibid.
117 Ibid. The council, having left the committee of the whole, approved the proposal on January 7, 1960.
118 Loo, "High Modernism," 46.
119 R.R. Logie to MMRA, August 4, 1960, LAC, RG 23, accession 1998-01421-5, box 13, file 5903-5.
120 Parker to Logie, August 1960, LAC, RG 23, accession 1998-01421-5, box 13, file 5903-5.
121 A.L. Pritchard to Logie, September 7, 1960, LAC, RG 23, accession 1998-01421-5, box 13, file 5903-5.
122 R.L. Butler, Petitcodiac River Report: Proposed Causeway to Cross at Moncton, December 19, 1963. Document received from Department of Fisheries and Oceans in response to Access to Information and Privacy (ATIP) request A-2014-00830.

123 J.L. Hart to L.E. Baker (Area Director, Department of Fisheries), January 9, 1961, LAC, RG 23, accession 1998-01421-5, box 13, file 5903-5.
124 Elson, Memo on Proposed Moncton Causeway Aboideau in Relation to Anadromous Fish of the Petitcodiac System, January 31, 1961, LAC, RG 23, accession 1998-01421-5, box 13, file 5903-5. Elson's research from the 1950s has been preserved on video: https://www.youtube.com/watch?v=0qngGFaiVUY.
125 MMRA, "Petitcodiac River Causeway Survey Report," March 30, 1961, 7, PANB, RS 126, Petitcodiac River Crossing.
126 Logie to Deputy Minister of Fisheries, May 17, 1961, LAC, RG 23, accession 1998-01421-5, box 13, file 5903-5.
127 Ibid; André Richard (Minister of Public Works) to J.A. Lévesque (Minister of Agriculture), May 28, 1961, NBDA.
128 MMRA, "Petitcodiac River Causeway Survey Report," March 30, 1961, PANB, RS 126.
129 New Brunswick Department of Public Works and Associated Designers and Inspectors, "Petitcodiac River Crossing Study," July-August 1961, 8, PANB, RS 126 (emphasis in original).
130 Ibid.
131 Town Planning Commission of Moncton, Some Comments on a Proposed Petitcodiac River Causeway, Fall 1961, PANB, RS 418, pt. 9 (emphasis in original).
132 Leonard Jones, Speech to Moncton Rotary Club, June 24, 1963, PANB, RS 418, pt. 9.
133 MMRA AC, July 7, 1964.
134 *Moncton Transcript,* June 13, 1966. This clipping came to my attention as part of a file provided to me by Nancy Hoar, who was active in the Lake Petitcodiac Preservation Association, which figures prominently in Chapter 4.
135 An engineer with the Department of Fisheries and Oceans, Chris Katopodis, explained: "In the vertical slot fishway, baffles are installed at regular intervals along the length to create a series of pools. Fish easily maintain their position within each pool, [while] travel between pools requires a burst of energy through each slot" (Katopodis, *Introduction to Fishway Design,* 3).
136 J.P. Parkinson (Chief, Resource Development Branch, Department of Fisheries, Atlantic Region) to E.F. Wheaton (Fish and Game Protective Association), April 11, 1967, LAC, RG 23, accession 1998-01421-5, box 13, file 5903-5.
137 The causeway closed off the river on February 10, 1968. A few days later, in an unrelated development, students at the Université de Moncton went on strike, taking to the streets in an unprecedented expression of Acadian impatience with their position as second-class citizens. The strike also included their efforts to speak French at a city council meeting, where they locked horns with Mayor Leonard Jones, a strong opponent of bilingualism. In a single week, the city experienced transformative changes to both its physical and political landscape. For more on the student strike, see Joel Belliveau, *In the Spirit of '68.* This moment has also been immortalized in film via Michel Brault and Pierre Perrault, *L'Acadie, L'Acadie?!?* (National Film Board of Canada, 1971).
138 R.L. Butler, "1968 Petitcodiac River Estuary: Causeway – Dam – Fishway," January 21, 1969, LAC, RG 23, accession 1998-01421-5, box 13, file 5903-5.
139 Ibid.
140 *Moncton Daily Times,* April 11, 1969.
141 Beaulieu, *Report on the Petitcodiac River Causeway,* 4. This situation is also discussed in Dominy *Petitcodiac River Causeway.*

142 "Report of Review Committee on Options for the Future of the Petitcodiac River Dam and Causeway," May 14, 1991, 5, PANB, RS 626, file 63635.
143 Minutes of City Council, October 17, 1968, City of Moncton Archives.
144 Butler, "1968 Petitcodiac River Estuary: Causeway – Dam – Fishway," January 21, 1969, LAC, RG 23, accession 1998-01421-5, box 13, file 5903-5.
145 MMRA AC, October 8–9, 1969.
146 Banks, *Preliminary Report on Tidal River Siltation*, 6.
147 Matas, "N.B. River Leads Endangered List."
148 Interview with Robert Ettinger, Windsor, NS, April 18, 2019.
149 MMRA, Preliminary Study, Avon River Causeway, 1965, LAC, RG 23, accession 1998-01421-5, box 17, file 5903-82-2-9.
150 Conlon to K.K. Dinnock (Mayor of Windsor), April 15, 1965, LAC, RG 23, accession 1998-01421-5, box 17, file 5903-82-2-9.
151 Interview with Bob Wilson, Falmouth, NS, April 17, 2019.
152 MMRA AC, June 9, 1965, app. A. Calculating costs over thirty-five years, there would be a benefit of $8.70 for every dollar invested in upkeep of the "present protective works," and only $5.80 for each dollar put into a "major tidal dam-causeway."
153 Ibid.
154 Matte to Sauvé, August 6, 1965, LAC, Maurice Sauvé Fonds, MG 32, file 10.
155 Buckley and Tihanyi, *Canadian Policies for Rural Adjustment*, 86. This study is discussed in Chapter 2.
156 Sylvain Cloutier (Assistant Secretary, Treasury Board) to L.E. Couillard (Deputy Minister of Forestry and Rural Development), January 10, 1968, LAC, Maurice Sauvé Fonds, MG 32, file 10.
157 Rural Development Branch, Department of Forestry and Rural Development, "Report on the MMRA program Arising out of Observations Made in Special Study No 7 by the Economic Council of Canada entitled Canadian Policies for Rural Adjustment by Buckley and Tihanyi, 1967," February 1968, PANB, RS 124, G39b.
158 Smith to C.P. Ruggles (Chief Biologist), May 11, 1965, LAC, RG 23, accession 1998-01421-5, box 17, file 5903-82-2-9.
159 Ibid.
160 Ibid.
161 MacEachern to A.W. Wigglesworth, May 19, 1965, LAC, RG 23, accession 1998-01421-5, box 17, file 5903-82-2-9.
162 The Fisheries Act that was in force in 1968 read as follows in terms of fishways: "Every slide, dam, or other obstruction across or in any stream where the Minister determines it to be in the public interest that a fish-pass should exist, shall be provided by the owner or occupier with a durable and efficient fishway or canal around the slide, dam, or obstruction" (SC 1932, c. 119, s. 20). It was left to the minister's discretion to determine what constituted "the public interest." Similarly, the regulations in force left the minister the option of acting if it was deemed appropriate, stipulating that the minister "may" intervene in the case of an obstruction, but not indicating that the minister "must" act (Canada, *Statutory Orders and Regulations*, 2: 1440).
163 H. Edwards to C.P. Ruggles, February 26, 1968, LAC, RG 23, accession 1998-01421-5, box 17, file 5903-82-2-9. Pesaquid was the Mi'kmaq name for the settlement at the convergence of the Avon and St. Croix Rivers, where the Town of Windsor (and the causeway) are now located. There are various spellings of the name for the "lake," but I am using the one that

was adopted by both the Nova Scotia Department of Transportation and the Windsor Town Council. Personal communications with Anna Allen, Mayor of Windsor, January 29, 2020.

164 Neil MacEachern, Memorandum. Re: Avon River Causeway, November 1, 1968, Department of Fisheries and Oceans (DFO), Dartmouth, NS, Avon Causeway File, as referenced in Isaacman, "Historic Examination of the Changes in Diadromous Fish Populations," 35. I tried to find this document in the DFO files in Dartmouth in the spring of 2019, but it was no longer in any of the pertinent folders. Grilse are Atlantic salmon that have spent one winter in the sea and have returned to fresh water to spawn.
165 Isaacman, "Historic Examination of the Changes in Diadromous Fish Populations," 35, 50. She maintained her respondents' anonymity.
166 Ibid., 78.
167 Town of Windsor, Council Minutes, April 22, 1968, Town of Windsor Archives.
168 Interview with Bob Wilson, Falmouth, NS, April 17, 2019.
169 Thurston, *Tidal Life*, 43.
170 *Hants Journal*, June 17, 1970; Nova Scotia Department of Highways, "Fact Sheet, Avon River Causeway-Dam," June 11, 1970, LAC, RG 124, accession 1984-85/617, box 6, file 1050-9.
171 Interview with Robert Ettinger, Windsor, NS, April 18, 2019.
172 *Hants Journal*, June 17, 1970.
173 J.A. Roberts to R.D. Gilbert (Deputy Minister of Agriculture), February 21, 1962, PANB, RS 124, F1M1.

Chapter 4: Legacies

1 Ouhilal, "Flushed," 100.
2 This figure includes the large and growing population of the Moncton region. Census figures show the population of its metropolitan area increasing from roughly 87,000 in 1971 (shortly after completion of the causeway) to nearly 140,000 in 2011 (just after opening of the gates).
3 McLaughlin, "Green Shoots," 151.
4 Hatvany, "Wetlands and Reclamation," 14: 274.
5 Ouhilal, "Flushed," 100. In addition to the record-breaking ride, there were various surfing exhibitions on the Petitcodiac over the course of a week. Ouhilal joined Whitbread, Wessels, and the French surfer Antony Colas, with an estimated 30,000 people watching.
6 CBC New Brunswick, July 25, 2013, https://www.cbc.ca/news/canada/new-brunswick/surfers-set-record-after-29-km-ride-on-moncton-tidal-bore-1.1327888. The link leads to a video of the surfers.
7 MMRA, *Marshland Utilization Statistics for the Year 1964*. This issue was discussed more fully in Chapter 2.
8 Jackson and Maxwell, *Landowners and Land Use in the Tantramar Area*, 3, 23.
9 Parson, "Regional Trends of Agricultural Restructuring," 346, 350.
10 More recently, the term *tintamarre* has become associated with the loud Acadian celebrations on August 15, the date of the Acadian national holiday. This practice began in 1955 on the bicentenary of the Acadian deportation, as a people who had long remained quiet made their public presence known. I have written about this first *tintamarre* in *Remembering and Forgetting in Acadie*, 63–65, 209–14.

11 Jackson and Maxwell, *Landowners and Land Use in the Tantramar Area*, 23, 26–27.
12 Burnett, *Passion for Wildlife*, 152. ARDA (the Agricultural Rehabilitation and Development Administration) was discussed in Chapter 2 as the agency that absorbed the staff of the MMRA when it closed, using it to assess projects that touched on the control of water across Atlantic Canada.
13 "Canada's National Wildlife Policy and Program," tabled in House of Commons, April 6, 1966, as cited in Burnett, *Passion for Wildlife*, 155.
14 Ibid., 64.
15 Thurston, *Tidal Life*, 86.
16 Burnett, *Passion for Wildlife*, 157.
17 Watson, *Pilot Project*, 1–2.
18 Green, "Chignecto Marshes," 46–47.
19 The Boyer slides were provided to me by Al Smith, who was involved with the acquisition of land in the Tintamarre National Wildlife area, from his position (starting in 1967) as a biologist working in the CWS office in Sackville, New Brunswick. His help – in various regards – was invaluable to both writing this book and producing *Unnatural Landscapes*.
20 Burnett, *Passion for Wildlife*, 151.
21 W.T. Munro, "The Role of National Wildlife Areas in the Maintenance and Distribution of Local Waterfowl Populations," in *The Waterfowl Habitat Management Symposium, Moncton, July 30–August 1, 1973* (n.p., n.d.), 30.
22 Green, "Chignecto Marshes," 48; Jackson and Maxwell, *Landowners and Land Use in the Tantramar Area*, 27; Burnett, *Passion for Wildlife*, 151. Burnett does not specify where Boyer wanted to stop the MMRA's intervention; this location was provided to me by Al Smith. For more on the Tintamarre NWA, see Environment and Climate Change Canada, Canadian Wildlife Service, Atlantic Region, *Tintamarre National Wildlife Area Management Plan* (Sackville, NB, 2016).
23 Green, "Chignecto Marshes," 48.
24 Whitman to J. Bryant, February 17, 1967, LAC, RG 109, file 498/04.
25 Treasury Board, file 681708, October 7, 1968, LAC, RG 109, file 498/04.
26 MMRA AC, March 26, 1955.
27 Environment and Climate Change Canada, Canadian Wildlife Service, Atlantic Region, *Shepody National Wildlife Area Management Plan* (Sackville, NB, 2018), 34.
28 American financial support for Ducks Unlimited Canada activities was visible on a plaque that I came across at the site of a water control structure it had constructed in 1970 for the Beach Pond Project in the Tintamarre NWA. The plaque indicated that the structure had been built by DUC "with funds supplied by Dr. John Newbold Robinson and his wife, Mrs. Betty Holland Robinson, of Wye Cottage Farm, Easton, Maryland, U.S.A., outstanding conservationists."
29 Hatvany, "Imagining Duckland," 228. In this article, Hatvany makes the case that DUC's efforts to construct a cross-border space for protection of waterfowl was evidence of a certain postnationalism, but it could also be read as an American effort to obscure the border, in the process advancing what Tina Loo called a form of "re-colonization." Loo, *States of Nature*, 190. There was nothing inconsistent about the DUC's actions being driven by both ecological thinking and an effort at recolonization. Hatvany's view is also advanced by Rod Giblett, who referred to this postnational space as "Duckland" in his book *Canadian Wetlands*. Giblett also presented this perspective in his *Postmodern Wetlands*.
30 Loo, *States of Nature*, 187.

31 Whitman, *Impoundments for Waterfowl*, 7–12.
32 Batt, *Marsh Keepers' Journey*, 177.
33 Ibid., 95; Leitch, *Ducks and Men*, 187. When Waugh made the move from the MMRA to DUC, he was joined by "Donald J. (Doc) Black, who had been John Waugh's right-hand man with MMRA and knew the marshes almost as well as John." Leitch, *Ducks and Men*, 187.
34 Interview with Al Smith, Sackville, NB, September 26, 2019.
35 Data from Environment and Climate Change Canada, Canadian Wildlife Service, Atlantic Region, *Management Plans* for the Tintamarre, Shepody, John Lusby, and Chignecto NWAs.
36 McLaughlin to Kim Jardine (New Brunswick Minister of Environment), November 16, 2001, PANB, RS 1034, 55691. One of the longest environmental controversies in Canadian history, the debate over the fate of the Petitcodiac Causeway has not been given any extended analysis, based on a wide range of sources. It has led to some excellent journalism, such as Mark Reid's "A River Divided," and has been the subject of Monique LeBlanc's very interesting documentary film *Petitcodiac* (Cinimage Productions, 2011). These treatments leaned largely on interviews with the main actors in the story, while the pages that follow – in addition to interviews with some of the same protagonists – are also grounded in the thousands of pages of archival documents held by both the federal and New Brunswick governments.
37 In addition to practicing as a gynecologist, Dr. McLaughlin also served in a number of senior administrative positions within the Canadian medical community. For this service, he received the Canadian Medical Association's Medal of Service in 2002, the first New Brunswick physician to be so honoured.
38 The expression was popularized at the time by Pete Seeger in the song *Little Boxes*, written in 1962 by Malvina Reynolds. It derisively referred to monotonous postwar suburban housing as having been made of "ticky-tacky." They were "Little boxes, all the same."
39 When I started the project, after listening to various people talk about Lake Petitcodiac, I looked for it on a map, where, of course, it could not be found.
40 Victor McLaughlin, submission to Riverview Town Council, July 10, 1995. Documents received from the Town of Riverview in response to an Access to Information and Privacy request.
41 Victor McLaughlin, "A Place in the Sun," June 25, 1989, written for CBC contest and found in documents provided by Nancy Hoar, Moncton, NB.
42 Dorothy and Don Murray to Joan Kingston (Minister of Environment), August 16, 1992, PANB, RS 269, 44394, vol. 1.
43 Ron and Darilyn Hill to Shawn Graham, November 12, 2007, PANB, RS 1075, 63155.
44 Buck to Sheldon Lee, September 15, 1992, PANB, RS 626, 63636. The sum of $150,000 in 1992 translates to nearly $250,000 in 2020, over 50 percent greater than the $160,000 average reported by the Canadian Mortgage and Housing Corporation for the resale of Moncton properties.
45 The figure is drawn from Locke and Bernier, *Annotated Bibliography*. Locke and Bernier presented 251 studies of aquatic life in the Petitcodiac watershed. Of these, 158 were published between 1968 (closure of the river) and 2000 (publication of their study). A number of the studies published in the earlier part of the 1960s were referenced in the previous chapter as the Petitcodiac project was being studied.

Notes to pages 182–89   259

46 Report from Conrad Bleakney, December 15, 1967, cited in letter from Gerald Tingley to Sheldon Lee, September 17, 1990, PANB, RS 748/5270-F-P, box 0-90-32/18988.
47 Wayne Kaye to Gerald Tingley, May 19, 1990, PANB, RS 314, 56649.
48 Conrad Bleakney reports: December 15, 1968, December 14, 1969, cited in letter from Tingley to Lee, September 17, 1990, PANB, RS748/5270-F-P, box 0-90-32/18988.
49 Bleakney to J.W. Bird (New Brunswick Minister of Natural Resources), September 20, 1979, PANB, RS 626, 63635 (emphasis in original).
50 Gary Griffin to W. Bishop (Minister of Transportation), September 25, 1979; Bleakney, "Letter to Editor," 1979, PANB, RS 626, 163-32.
51 Semple, *Anadromous Fish Stocks,* 27.
52 ADI Ltd., *Study of Operational Problems, Petitcodiac River Causeway,* December 1979, PANB, RS 24, 67028; Hans Jensen to John Ritter, General Comments re Petitcodiac Causeway, February 22, 1991, document received from DFO in response to ATIP request A-2014-00830.
53 Petitcodiac Riverkeeper, *Response to First Draft of Special Advisor, Petitcodiac Causeway Review,* January 31, 2001, 6, appended to Niles, *Review of the Petitcodiac Causeway and Fish Passage Issues.* Taking a similar view, in 1980 DFO officials concluded that "fish passage facilities in general do not work well on tidal rivers and that they were not confident that any artificial fish passage facilities will restore salmon runs at this site" (cited in Petitcodiac Causeway, Comments, February 21, 1980, PANB, RS 626, 163-32).
54 Barron to Wilfrid Bishop (Minister of Transportation), February 22, 1980, PANB, RS 626, 163-32.
55 *Moncton Transcript,* September 30, 1980.
56 Pettigrew to E.T. Owens, October 3, 1980, PANB, RS 106, 58409.
57 *Moncton Transcript,* September 30, 1980.
58 Pettigrew to E.T. Owens, October 3, 1980, PANB, RS 106, 58409.
59 Gilchrist to G. Del Reeleder (Deputy Minister of Transportation), February 5, 1980, PANB, RS 626, 163-32.
60 Ibid.
61 Buck to Sheldon Lee, September 15, 1992, PANB, RS 626, 63636.
62 Inter-Departmental Committee Report on Options for the Future of the Petitcodiac River Dam and Causeway, May 14, 1991, 16, PANB, RS 626, 63635. Smolts are young salmon, ready to migrate to salt water.
63 Harvey, "Death Watch on the Petitcodiac," 39–40.
64 Victor LeBlanc, testimony in documentary film *Petitcodiac* (Cinimage, 2011).
65 Gary Griffin, Petitcodiac River Restoration Proposal presented to Morris Green, Minister of Natural Resources and Energy, February 1988, PANB, RS 748/5270-F-P, box 0-90-32/18988.
66 DFO, ADM Report, Freshwater and Anadromous Division, Opening of the Tidal Gates at Petitcodiac River Causeway, May 1988, document received from DFO in response to ATIP request A-2014-00830.
67 Harvey, *Death Watch,* 41.
68 Betts to Hubert Seamans (Chair of Lake Petitcodiac Study Committee), November 4, 1988, PANB, RS 418, 65541.
69 *Times & Transcript,* June 14, 1988.
70 Betts to Hubert Seamans, November 4, 1988, PANB, RS 418, 65541.

71 J.A. Ritter, "Effects of Anadromous Fish Stocks of the Petitcodiac River," March 1991, PANB, RS 314, 56649. Similarly, a study by the firm ADI (which had earlier rejected opening of the gates) found that the largest number of people would benefit from opening the gates. ADI, "Analysis of options for the future of the Petitcodiac River Dam and Causeway," May 1992, PANB, RS 269, 44393.
72 Victor LeBlanc to Sheldon Lee, September 17, 1991, PANB, RS 626, 63636 (emphasis in original).
73 Mitton to McKenna, June 19, 1991, PANB, RS 748/5270-1.
74 Bird, manuscript letter to editor (no newspaper noted), August 20, 1991, PANB, RS 748/5270-1.
75 *Times & Transcript,* September 28, 1991.
76 Griffin to McKenna, October 10, 1991, PANB, RS 748/5270-1.
77 Gary Griffin, testimony before public hearing of Round Table on Environment and Economy, Moncton, November 30 – December 1, 1989, PANB, RS 967/47028.
78 Premier's Round Table on Environment and Economy, *Towards Sustainable Development in New Brunswick,* 13; Premier's Round Table on Environment and Economy, *Because We Want to Stay,* 77, 79.
79 Executive Council: Cabinet Meeting Records, June 25, 1992, PANB, RS 9.
80 Department of Environment Backgrounder, November 1992, PANB, RS 967/47028.
81 Julia Chadwick and Gary Griffin (Friends of the Petitcodiac), n.d., text found in documents provided by Nancy Hoar, Moncton, NB.
82 Griffin to McKenna, June 30, 1992, PANB, RS 748/5270-F-P, box 0-92-40.
83 Douglas Hamer to Sheldon Lee, August 19, 1992, PANB, RS 626, 63636. Hamer included beauty shots of the headpond to support his letter.
84 Gary and Jane Sherrard to Sheldon Lee, September 11, 1992, PANB, RS 626, 63636.
85 Hans Jensen to John Ritter, General Comments re Petitcodiac Causeway, February 22, 1991, document received from DFO in response to ATIP request A-2014-00830.
86 Memorandum to the Executive Council, November 30, 1992, NBDA.
87 Susan LeBlanc to McKenna, October 1, 1992, PANB, RS 748/5270-F-P, box 0-92-40. While I have generally translated French quotes into English, here I have retained the original in the text to underscore the Acadian connection. The quote reads in English: They sought reversal of "a decision which permits the continued destruction of the Petitcodiac watershed."
88 This film is discussed at length in the Prologue.
89 Paul Surette, "History of the River's First Settlers," presented to Petitcodiac River Symposium, September 26, 1992, PANB, RS 269, 44393.
90 DesNeiges to Premier Camille Theriault, October 30, 1998, PANB, RS 1033. The original was in French. Born Michel LeBlanc, he changed his name to LeBlanc des Neiges in the 1990s, before simply going by Michel DesNeiges, which is how he is referenced in these pages.
91 Belliveau, *In the Spirit of '68;* Rudin, *Kouchibouguac.*
92 Interview with Daniel LeBlanc, Bouctouche, NB, September 25, 2019.
93 DesNeiges to Premier Camille Theriault, October 30, 1998, PANB, RS 1033. The original was in French.
94 Ibid.
95 *Times & Transcript,* November 28, 1992.

96 Crazy Horse was the Lakota leader in 1876 at what they refer to as the Battle of the Greasy Grass (also known as the Battle of Little Bighorn), while Louis Riel was the Métis leader of risings in 1869–70 (in present-day Manitoba) and 1885 (in present-day Saskatchewan), the second of which resulted in his hanging. Beausoleil Broussard was an Acadian who resisted British efforts to deport his people at the time of the *grand dérangement*.

97 The song was on the Zéro Degré Celsius album *Le feu sur la rue Main* (1995), and on Richard's album *Cap Enragé* (1996). A description of the song does not do it justice. It needs to be heard. Email correspondence with Yves Chiasson, July 31, 2019; Interview with Daniel LeBlanc, Bouctouche, September 25, 2019. Opposition was also mobilized through the 1995 song *Causeway, No Way*, recorded by the Moncton band Les Païens and written by Chris LeBlanc, one of its members. It was on their 1999 album EP Phonde. The lyrics focus on the buildup of silt: "Causeway, no way / Not in my hometown / Causeway, no way / The blood on our hands is brown."

98 Student (name of minor withheld) to Marcelle Mersereau, June 15, 1994, PANB, RS 314, 56649. The original was in French. Michel DesNeiges indicated to me that the people who went into the schools were students from the Université de Moncton connected with Écoversité, the group he had helped found. Interview with Michel DesNeiges, September 25, 2019.

99 Student (name of minor withheld) to Marcelle Mersereau, June 20, 1994, PANB, RS 314, 56649. The original was in French.

100 Student (name of minor withheld) to Marcelle Mersereau, January 14, 1997, PANB, RS 748/5270-P/0-97-45, 19375. Students were mobilized once again in 2005, when a teacher at École Ste-Bernadette in Moncton organized transmission of a petition to both Prime Minister Paul Martin and Premier Bernard Lord. Isabelle-Andrée Lang to Paul Martin et al., April 14, 2005, PANB, RS 1034, 55827.

101 Memorandum to the Executive Council, November 30, 1992, NBDA (emphasis in original).

102 Frank McKenna, Speaking notes for memo to Cabinet, Petitcodiac Causeway Gates, December 8, 1992, PANB, RS 269.

103 Alyre Chiasson, "A Flow Control Model for the Petitcodiac River," November 1994, PANB, RS 314, 56649; DFO, Briefing Notes for the Minister, September 26, 1996, document received from DFO in response to ATIP request A-2014-00830.

104 *Telegraph-Journal*, January 28, 1997.

105 Ibid.

106 *Times & Transcript*, November 16, 1995.

107 *LAPPA vs. Canada (Minister of the Environment)*, Federal Court of Canada, File T-1132-98, June 9, 1998.

108 Ed Baxter to Minister of the Environment, August 23, 1997, PANB, RS 269, 44394.

109 Mary and Blair Dolan to James Lockyer (MLA), January 2, 1997, PANB, RS 269, 44395.

110 Ronald and Beryl Gaskins to Minister of the Environment, November 8, 1995, PANB, RS 269, 44395.

111 Suzanne Doucet to McKenna, n.d., PANB, RS 269, 44395. The original was in French.

112 Reisner, *Cadillac Desert*, 511–14.

113 Ibid., 513.

114 Cronin and Kennedy, *The Riverkeepers*, 18–19.

115 A few years earlier, Kennedy had played an important role in another Canadian context, supporting the ultimately successful challenge by the Cree of northern Quebec to stop the development of Hydro-Québec's Great Whale project. This story is compellingly told in Magnus Isaacson's film *Power* (Montreal: NFB, 1996).

116 *Telegraph-Journal*, February 25, 1999; Interview with Michel DesNeiges, Moncton, September 25, 2019.

117 I discuss this effort at international recognition of the deportation in *Remembering and Forgetting in Acadie*, 22–34. LeBlanc's role as Petitcodiac Riverkeeper intersected with his involvement with the commemoration of the deportation, when in 2005 (on the 250th anniversary of the *grand dérangement*) the commission inaugurated its first monument on the banks of the Petitcodiac, five kilometres downstream from the causeway.

118 LeBlanc, *Petitcodiac;* Interview (recorded in French) with Daniel LeBlanc, Bouctouche, September 25, 2019.

119 Niles, *Review of the Petitcodiac Causeway and Fish Passage Issues.*

120 McLaughlin to Niles, January 27, 2001, PANB, RS 839, 68491.

121 LeBlanc to Herb Dhaliwal, September 25, 2001, PANB, RS 1034, 55691.

122 This distinction was bestowed in 2003 by Earthwild International and Wildcanada.net. In 2002, when the list of endangered rivers was first published, the Petitcodiac was included as the second most endangered river in Canada (http://petitcodiac.org/river-restoration-campaign-1999-2010/).

123 *Times & Transcript*, January 1, 2002.

124 AMEC Earth and Environmental, *Environmental Impact Assessment Report*, 363–64.

125 Ibid., 365–66.

126 Petitcodiac Riverkeeper, Press Release, October 25, 2005, PANB, RS 1034, 55827.

127 New Brunswick, Department of Environment, Verbatim Transcript of the Public Meeting held on 29 November 2005 to Discuss the Environmental Impact Assessment for the Modifications to the Petitcodiac River Causeway, document found in collection provided by Nancy Hoar, Moncton, NB. Jim Sellars expanded on these views in a self-published treatment, completed in 2009, that he provided to me. In *There Is Money in Here Somewhere*, he continues along the same lines as the intervention cited here, questioning the science that led to the decision to open the gates, and figuring that everyone was padding their pockets along the way. In the book, he makes numerous negative references to the University of Moncton (always in the English form), at one point referring to its students – who supported opening the gates – as "'true believers of the faith' without a notion of sense among them." As we have seen, delegitimizing expertise at the Acadian university was standard fare in the Petitcodiac controversy.

128 This figure is from Daniel LeBlanc, "Studies Show the Petitcodiac Will Largely Restore Itself."

129 New Brunswick, Department of Environment, Verbatim Transcript of the Public Meeting held on 29 November 2005 to Discuss the Environmental Impact Assessment for the Modifications to the Petitcodiac River Causeway, document found in collection provided by Nancy Hoar, Moncton, NB.

130 Petitcodiac Riverkeeper, Press Release, July 6, 2007, http://nben.ca.

131 Terry Dolan to Shawn Graham, August 7, 2007, PANB, RS 1075, 63155.

132 Jerry O'Rourke to Shawn Graham, April 16, 2010, PANB, RS 1075, 63155. McKenna's retreat was discussed earlier in the chapter.

133 Interview with Jim Reicker, Boundary Creek, NB, September 25, 2019.

134 Julia Chadwick and Gary Griffin (Friends of the Petitcodiac), n.d., text found in documents provided by Nancy Hoar, Moncton, NB.
135 Niles, *Review of the Petitcodiac Causeway and Fish Passage Issues*, 29–30.
136 Interview with Wiebe Leenstra, Boundary Creek, NB, September 24, 2019. The Leenstra family emigrated from the Netherlands in the 1990s, so they came with experience in dealing with dykeland.
137 James Reicker to Office of Ombudsman, October 31, 2009, PANB, RS 1075, 63219. The legislation that gave the province access to Reicker's property was *An Act to Amend the Public Works Act*, SNB 2009, c. 1.
138 *Lake Petitcodiac Preservation Association Inc. c. NB Province*, 2010, 360 RNB (2d) 178.
139 *Times & Transcript*, April 15, 2010.
140 Ibid., May 4, 2010.
141 Petitcodiac Watershed Alliance, "Five Years of Study on Fish Populations Since Opening of the Causeway Gates Show Signs Petitcodiac River Ecosystem Health Improving" (news release, June 15, 2015), http://ffhr.ca/wp-content/uploads/2015/04/News-Release-Petitcodiac-Fish-Recovery-Coalition-June20151.pdf. Also see a description of Redfield's work at http://petitcodiac.org/my-river/.
142 Petitcodiac Watershed Alliance, "Five Years of Study on Fish Populations."
143 Colas, *Mascaret*, 160; Ouhilal, "Flushed," 94. Colas's text was originally in French.
144 Colas, *Mascaret*, 162. Among the surfers was Melvin Perez, a bartender at the Château Moncton, the hotel where the Americans were staying. Perez had been a surfer in his native Costa Rica, and so decided to go out with the others. He tells his story at "Melvin Perez Proud to Be a Petitcodiac Tidal Bore Surfing Pioneer," Petitcodiac Riverkeeper, February 14, 2014, http://petitcodiac.org/melvin/.
145 Interview with Daniel LeBlanc, Bouctouche, NB, September 25, 2019.
146 Arsenault and Ouellette, "La restauration de la rivière Petitcodiac," 41–42.
147 Gary Griffin, manuscript letter to editor, June 1991, PANB, RS 748/5270-1.
148 Van Proosdij, *Intertidal Morphodynamics*, 43–46. She figures prominently, narrating the Windsor Salt Marsh, in the final segment of the film *Unnatural Landscapes*.
149 "Monitoring the Health of the Avon River Estuary," *TwinNews*, September 2018, https://hwy101windsor.ca/wp-content/uploads/2018/09/Vol2_Issue1_101Twinning-Newsletter-September-2018-Final.pdf; Daborn et al., *Ecological Studies of the Windsor Causeway and Pesaquid Lake, 2002*, 58.
150 Van Proosdij and Townsend, "Sedimentation and Mechanisms of Salt Marsh Colonization," 261.
151 Daborn et al., *Environmental Implications of Expanding the Windsor Causeway*, 12.
152 Szabo-Jones, "Secrets of the Salt Marsh." The webcam is at https://www.novascotiawebcams.com/en/webcams/windsor-salt-marsh. On the issue of sewage, see Daborn et al., *Ecological Studies of the Windsor Causeway and Pesaquid Lake, 2002*, 58.
153 *Hants Journal*, May 21, 1986.
154 Conrad and Semple, *Fish Passage at Avon River Causeway*.
155 The Pumpkin Regatta is possible only because of the nearby farm of Howard Dill, which grows the largest pumpkins in the world. For more on the regatta, visit http://worldsbiggestpumpkins.com.
156 Nova Scotia, Department of Transportation and Infrastructure Renewal, "Highway 101 Twinning and Avon River Aboiteau and Causeway, Project Update, February 27, 2019," https://hwy101windsor.ca/wp-content/uploads/2019/02/CLC-Meeting-Presentation-2019-02-27.pdf.

157 Friends of the Avon River, "The Avon River Causeway, Windsor, Nova Scotia Is One of Canada's Biggest Man-Made Disasters," http://www.viewzone.com/avonriver.html.
158 *Hants Journal,* July 23, 2003.
159 *Valley Journal-Advertiser,* July 10, 2017.
160 *Kings County Advertiser,* October 18, 2018.
161 CLC Meeting Presentation, February 27, 2019, https://hwy101windsor.ca. Many thanks to Robert Buranello and Astrid Friedrich for providing this photo from their Windsor storefront.
162 E-petition, E-2182, initiated by Sonja Wood, May 14, 2019, https://petitions.ourcommons.ca/en/Petition/Details?Petition=e-2182; CLC Meeting Presentation, February 27, 2019.
163 Interview with Gary Griffin, Moncton, NB, May 14, 2015.
164 Interview with Daniel LeBlanc, Bouctouche, NB, September 25, 2019.

Epilogue: Meet the Grand Pré Marsh Body

1 In 1952, 119 marsh bodies were up and running, having been incorporated since the introduction of the new regime in the late 1940s.
2 Interview with Robert Palmeter recorded for *Unnatural Landscapes,* Grand-Pré, NS, June 2018.
3 Greenberg et al., "Climate Change, Mean Sea Level and High Tides in the Bay of Fundy." The sinking of the land mass is discussed in Henton et al., "Crustal Motion and Deformation Monitoring of the Canadian Landmass," 180.
4 Falmouth Great Dyke Marsh Body, Minutes (private collection, Philip Davison).
5 For the significance of this site of memory, see Rudin, *Remembering and Forgetting in Acadie,* pt. 2.
6 Campbell, *Nature, Place, and Story,* 54–55.
7 CTV News Atlantic, "New Dykeland Awareness Signs Raising Eyebrows in Grand Pre," July 23, 2015, https://atlantic.ctvnews.ca/new-dykeland-awareness-signs-raising-eyebrows-in-grand-pre-1.2483229.
8 Campbell, *Nature, Place, and Story,* 54–55.
9 Ibid., 67.
10 Nomination Grand Pré, *World Heritage Nomination Proposal,* 31; Wicken, "Re-examining Mi'kmaq-Acadian Relations," 100.
11 Interview with Robert Palmeter recorded for *Unnatural Landscapes,* Grand-Pré, NS, June 2018.
12 Interviews recorded for *Unnatural Landscapes* with Ernie Partridge, Dorchester, NB, and Ann Rogers, Riverview, NB, June 2018.
13 Canada, Parliament, Senate, Standing Special Committee on Land Use, April 26, 1961.
14 Scott, *Seeing Like a State,* 340.

Bibliography

ARCHIVAL SOURCES

Acadia University Archives
George Frail Fonds (2015.035-FRA)
Grand Pre Marsh Body Fonds (2007.032; 2010.047)
MMRA Fonds (1998.001/1/2)

Annapolis Heritage Society (O'Dell House Museum, Annapolis Royal, NS)
George Frail Collection

Canada, Department of Agriculture, Nappan Experimental Farm (Nappan, NS)
Daily Journal

Canada, Department of Fisheries and Oceans (Dartmouth, NS)
Avon Causeway File
Avon River Fishway File (NS 61)

King's County Museum (Kentville, NS)
Minutes of Grand Pre Common Field (MS11)

Library and Archives Canada (LAC)
Canadian Wildlife Service Records (RG 109)
Department of Agriculture Records (RG 17)
Department of Finance Records (RG 19)
Department of Fisheries Records (RG 23)
Department of Regional Economic Expansion Records (RG 124)
Maurice Sauvé Fonds (MG 32)
Privy Council Records (RG 2)

Mount Allison University Archives
Albert Anderson Interview (9825)
Donald Harper Fonds (7740)
Robert J. Cunningham Fonds (7922)

National Film Board of Canada Archives (Montreal)
Production file for *Les aboiteaux* (53-144)

New Brunswick Department of Agriculture (NBDA)
Uncatalogued documents held at Moncton offices

Nova Scotia Archives
Royal Commission on Provincial Development and Rehabilitation Fonds

Nova Scotia Department of Agriculture (NSDA)
Uncatalogued documents held at offices in Nappan, Kentville, Truro, and Windsor

Provincial Archives of New Brunswick (PANB)
Bernard Lord Papers (RS 1034)
Camille Theriault Papers (RS 1033)
Department of Agriculture, Fisheries and Aquaculture Records (RS 123/RS 124)
Department of Environment and Local Government (RS 967)
Department of Municipal Affairs and Environment Records (RS 269)
Department of Public Works Records (RS 126)
Department of Transportation Records (RS 626)
Deputy Minister of Natural Resources Records (RS 106)
Deputy Minister of Supply and Services Records (RS 693)
Environmental Planning and Sciences Branch Records (RS 839)
Environmental Protection Records (RS 314)
Executive Council Records (RS 9)
Frank McKenna Papers (RS 748)
Legislative Assembly: Sessional Records (RS 24)
Moncton Municipal Records (RS 418)
Shawn Graham Papers (RS 1075)

Saskatchewan Archives
James G. Gardiner Fonds (F 65, R-1022)

Town of Windsor Archives (NS)
Council Minutes, 1964–70

PRIVATE COLLECTIONS

Boyer, Joe. Slide collection. In the possession of Al Smith (Sackville, NB).
Desplanque, Constant. "The Story of My Life" (unpublished autobiography). In the possession of David Mossman (Wolfville, NS).
Hoar, Nancy (Moncton, NB). Collection of documents dealing with the Petitcodiac Causeway.

Bibliography

Marsh Body Minute Books (with reference to the individual in possession of collection)
Bishop Beckwith (Eric Patterson)
Dentiballis (Andi Rierden)
Falmouth Great Dyke (Philip Davison)
Horton (Charles Curry)
Le Farm (Jim Inglis)
Queen Ann (Larry Hudson and Peter de Nuke)
Victoria Diamond Jubilee (Tara Hill)

Access-to-Information Requests

Canada, Department of Fisheries and Oceans. Documents Pertinent to the Petitcodiac Causeway, A-2014-00830.
City of Moncton, Moncton City Council Meetings, 1960–2010.
New Brunswick, Department of Transportation and Infrastructure. Documents Pertinent to the Petitcodiac Causeway, RTI-1920-9.
Town of Riverview, Documents Pertinent to the Petitcodiac Causeway.

Interviews by and Personal Communications with Author

Anna Allen
Roger Bacon
Jim Bremner
Alyre Chiasson
Graham Daborn
Michel DesNeiges
George Dorsay
Shannon Douthwright
Robert Ettinger
Gary Griffin
Dick Haliburton

Nancy Hoar
Hank Kolstee
Mary Laltoo
Daniel LeBlanc
Victor LeBlanc
Wiebe Leenstra
Robert Palmeter
Ernie Partridge
Arthur Phinney
James Reicker
Ann Rogers

Jim Sellars
Al Smith
Harry Thurston
George Trueman
Danika van Proosdij
John Waugh
Bob Wilson
John Wilson
Jim Wood
Sonja Wood

Video Sources

Les aboiteaux. Directed by Léonard Forest. Montreal: Office national du film, 1955. http://www.onf.ca/film/aboiteaux. Available with English subtitles: *The Dikes*, https://www.nfb.ca/film/dikes_en.
Petitcodiac. DVD. Directed by Monique LeBlanc. Cinimage Productions, 2011.
Petitcodiac. VHS. Directed by Paul Arseneau. Connections Productions, 1999.

Printed Primary Material

AMEC Earth and Environmental. *Environmental Impact Assessment Report for Modifications to the Petitcodiac River Causeway*. Moncton, 2005.
Baird, W.W. *Dominion Experimental Farm, Nappan, N.S., Progress Report, 1937–1947*. Ottawa: Department of Agriculture, 1947.

–. *Report on Dyke Reclamation 1913 to 1952.* Ottawa: Canada Department of Agriculture, Experimental Farm Service, 1954.

Banks, C.A. *Preliminary Report on Tidal River Siltation in the Bay of Fundy Area.* Amherst, NS: Soil and Water Division, Department of Forestry and Rural Development, 1969.

Beaulieu, G.T. *Report on the Petitcodiac River Causeway.* Halifax: Department of Fisheries and Forestry, 1970.

Boardman, Samuel. "Tidal Lands and Diked Marshes of Nova Scotia and New Brunswick." In *Tide Marshes of the United States,* edited by D.N. Nesbit, 33–51. Washington, DC: Government Printing Office, 1885.

Buckley, Helen, and Eva Tihanyi. *Canadian Policies for Rural Adjustment: A Study of the Economic Impact of ARDA, PFRA, and MMRA.* Study prepared for the Economic Council of Canada. Ottawa: Queen's Printer, 1967.

Canada. *Public Accounts for the Fiscal Year Ended March 31, 1943.* Ottawa: King's Printer, 1944.

–. *Report of the Minister of Agriculture for the Year Ended March 31, 1945.* Ottawa: King's Printer, 1945.

–. *Statutory Orders and Regulations, Consolidation 1955.* Ottawa: Queen's Printer, 1955.

Canada, Department of Agriculture. *Machines for Marshland Ditching.* Ottawa: Canada Department of Agriculture, 1963.

Canada, Public Health Engineering Division, Department of National Health and Welfare. *Aspects of Pollution, Annapolis River, Nova Scotia.* Ottawa: Queen's Printer, 1967.

Champlain, Samuel de. "Voyages of Samuel de Champlain." In *The Works of Samuel de Champlain,* translated by W.F. Ganong. Toronto: Champlain Society, 1922.

Le Congrès mondial acadien: L'Acadie en 2004, Actes des conférences et des tables rondes. Moncton: Éditions d'Acadie, 1996.

Conrad, Vern, and J.R. Semple. *Fish Passage at Avon River Causeway and Fish Enhancement Potential.* Halifax: Department of Fisheries and Oceans, 1987.

Dominy, C.L. *Petitcodiac River Causeway, Fishway Evaluation.* Manuscript Report 70-3. Halifax: Department of Fisheries, 1970.

Experimental Farm, Nappan, NS. *75th Anniversary: Progress through Research, 1867–1962.* Ottawa: Department of Agriculture, 1962.

Haase, Gordon. *Some Economic Aspects of Marshland Reclamation in the Maritime Provinces.* Ottawa: Canada Department of Agriculture, 1954.

Haase, Gordon, and D.J. Packman. *Marshland Utilization in Nova Scotia and New Brunswick.* Ottawa: Department of Agriculture, 1953.

Lattimer, J.E. *A Study of 128 Profitable Cumberland County Farms.* Halifax: Nova Scotia Department of Agriculture, 1943.

–. *What Farmers Told Us in Cumberland County.* Halifax: Nova Scotia Department of Agriculture, 1941.

Maritime Marshland Rehabilitation Administration (MMRA). *Annual Reports, 1948–70.* Ottawa: Government Printer.

–. *Marshland Utilization Statistics for the Year 1964.* Amherst, NS: MMRA, 1965.

McIntyre, R.R., and C. Desplanque. "Silt Deposition and Erosion in the Tidal Rivers of the Bay of Fundy." Paper presented to Royal Agricultural Society of the Commonwealth, Banff, October 1959.

McIntyre, R.R., and R.H. Palmeter, "Protection of Earth Structures by Vegetative Cover." *Canadian Agricultural Engineering* 4 (1962): 14–16.

Nesbit, D.N. *Tide Marshes of the United States.* Washington, DC: Government Printing Office, 1885.
Niles, Eugene. *Review of the Petitcodiac Causeway and Fish Passage Issues.* Halifax: Department of Fisheries and Oceans, 2001.
Partridge, Ernest J. *Memories.* Amherst, NS: Acadian Printing, 2019.
Premier's Round Table on Environment and Economy. *Because We Want to Stay: Sustainable Development in New Brunswick.* Fredericton: Province of New Brunswick, 1992.
–. *Towards Sustainable Development in New Brunswick.* Fredericton: Province of New Brunswick, 1992.
Report of the New Brunswick Committee on Reconstruction. Fredericton: Province of New Brunswick, 1944.
Sellars, Jim. *There Is Money in Here Somewhere: A Story of Government Corruption and Environmental Chicanery, and Brainless Mistakes That Keep Us from Getting to the Eden We Already Enjoy.* Moncton, ca. 2009.
Semple, J.R.. *Anadromous Fish Stocks in the Petitcodiac River System and the Moncton Causeway: A Status Report.* Halifax: Department of Fisheries and Oceans, 1979.
Sheldon, John. *To Canada, and through It, with the British Association.* Ottawa: Department of Agriculture, 1886.
Spence-Sales, Harold. *Guide to Urban Dispersal.* Montreal: McGill University, Committee on Physical Planning, 1956.
–. *Moncton Renewed.* Moncton: Urban Renewal Survey, 1958.
Summary Report on Project NB 22: Tantramar River Survey, Westmorland County, NB, February 15, 1953. Amherst, NS: MMRA, 1953.
Warren, George. *Tidal Marshes and Their Reclamation.* US Department of Agriculture, Office of Experiment Stations, Bulletin 240. Washington, DC: Government Printing Office, 1911.
Watson, G.H. *Pilot Project, Marsh Acquisition, Cumberland County, Nova Scotia.* Sackville, NB: Canadian Wildlife Service, 1965.
Whitman, W.R. *Impoundments for Waterfowl.* Occasional Paper No. 22. Ottawa: Canadian Wildlife Service, 1976.
Woods, Walter S. *Rehabilitation: Being a History of the Development and Carrying Out of a Plan for the Re-establishment of a Million Young Veterans of World War II by the Department of Veterans Affairs.* Ottawa: Queen's Printer, 1953.

SECONDARY SOURCES

Anstey, T.H. *One Hundred Harvests: Research Branch, Agriculture Canada, 1886–1986.* Ottawa: Agriculture Canada, Research Branch, 1986.
Archibald, E.S. "Maritime Dykeland Rehabilitation." *Canadian Geographic Journal* 28 (1944): 198–202.
Armstrong, Christopher, Matthew Evenden, and H.V. Nelles. *The River Returns: An Environmental History of the Bow.* Montreal and Kingston: McGill-Queen's University Press, 2009.
Arsenault, Gabriel, and Roger Ouellette, "La restauration de la rivière Petitcodiac (1968–2016): Une analyse basée sur l'approche des courants multiples." *Journal of New Brunswick Studies* 11 (2019): 31–47.

Batt, Bruce. *The Marsh Keepers' Journey: The Story of Ducks Unlimited Canada*. Winnipeg: Ducks Unlimited Canada, 2012.
Belliveau, Joel. *In the Spirit of '68: Youth Culture, the New Left, and the Reimagining of Acadia*. Translated by Käthe Roth. Vancouver: UBC Press, 2019.
Bleakney, J. Sherman. *Sods, Soils and Spades: The Acadians at Grand-Pré and Their Dykeland Legacy*. Montreal and Kingston: McGill-Queen's University Press, 2004.
Brown, Paul. "A Lesson in Bureaucratic Persistence: The Provision and Rehabilitation Management Services in the Maritimes, 1943–81." *Canadian Public Administration* 25 (1982): 130–46.
Burnett, J. Alexander. *A Passion for Wildlife: The History of the Canadian Wildlife Service*. Vancouver: UBC Press, 2003.
Butzer, Karl. "French Wetland Agriculture in Atlantic Canada and Its European Roots: Different Avenues to Historical Diffusion." *Annals of the Association of American Geographers* 92 (2002): 451–70.
Calder, Frank. *100 Years of Agricultural Research: The History of the Experimental Farm, Nappan, NS*. Kentville, NS: Agriculture Canada Research Branch, 1988.
Campbell, Claire. *Nature, Place, and Story: Rethinking Historic Sites in Canada*. Montreal and Kingston: McGill-Queen's University Press, 2017.
Campbell, Claire, and Robert Summerby-Murray, eds. *Land and Sea: Environmental History in Atlantic Canada*. Fredericton: Acadiensis Press, 2013.
Clark, Andrew Hill. *Acadia: The Geography of Early Nova Scotia to 1760*. Madison: University of Wisconsin Press, 1968.
Colas, Antony. *Mascaret: l'onde lunaire*. Anglet, France: YEP, 2014.
Cormier, Yves. *Les aboiteaux en Acadie: hier et aujourd'hui*. Moncton, NB: Chaire d'études acadiennes, 1990.
–. "Nos aboiteaux." *Les cahiers de la Société historique acadienne* 19, 1–2 (1988): 5–17.
Cronin, John, and Robert F. Kennedy Jr. *The Riverkeepers: Two Activists Fight to Reclaim Our Environment as a Basic Human Right*. New York: Scribner, 1997.
Cronon, William. *Nature's Metropolis: Chicago and the Great West*. New York: W.W. Norton, 1992.
Cunningham, R.J. "Thompson, Toler (Tolar)." In *Dictionary of Canadian Biography*, http://www.biographi.ca/en/bio/thompson_toler_7E.html.
Daborn, Graham, et al. *Ecological Studies of the Windsor Causeway and Pesaquid Lake, 2002*. Report Prepared for Nova Scotia Department of Transportation and Public Works. Publication No. 69. Wolfville: Acadia Centre for Estuarine Research, 2003.
–. *Environmental Implications of Expanding the Windsor Causeway*. Final Report Prepared for Nova Scotia Department of Transportation and Public Works. Publication No. 72. Wolfville: Acadia Centre for Estuarine Research, 2003.
Daigle, Jean. "Acadian Marshland Settlement." In *Historical Atlas of Canada*, vol. 1, edited by R. Cole Harris, Plate 29. Toronto: University of Toronto Press, 1987.
Dalton, F. Keith. "Fundy's Prodigious Tides and Petitcodiac's Tidal Bore." *Journal of Royal Astronomical Society of Canada* 45 (1951): 225–31.
Davidson, Nick C. "How Much Wetland Has the World Lost? Long-Term and Recent Trends in Global Wetland Area." *Marine and Freshwater Research* 65 (2014): 934–41.
Daye, Vera. "Booming, Bustling Moncton." *Atlantic Advocate* 48 (October 1957): 15–21.

Desplanque, C., and D.J. Mossman. "Tides and Their Seminal Impact on the Geology, Geography, History, and Socio-Economics of the Bay of Fundy, Eastern Canada." *Atlantic Geology* 40 (2004). https://doi.org/10.4138/729.
–. "Storm Tides of the Bay of Fundy." *Geographical Review* 89, 1 (1999): 23–33.
Dummitt, Christopher. *The Manly Modern: Masculinity in Postwar Canada*. Vancouver: UBC Press, 2007.
Faragher, John Mack. *A Great and Noble Scheme*. New York: W.W. Norton, 2005.
Fiege, Mark. *Irrigated Eden: The Making of an Agricultural Landscape in the American West*. Seattle: University of Washington Press, 1999.
–. "The Weedy West: Mobile Nature, Boundaries, and Common Space in the Montana Landscape." *Western Historical Quarterly* 36, 1 (Spring 2005): 22–47.
Fifty Years of Progress on Dominion Experimental Farms, 1886–1936. Ottawa: Canada Department of Agriculture, 1936.
Fitzgerald, Deborah. *Every Farm a Factory: The Industrial Ideal in American Agriculture*. New Haven, CT: Yale University Press, 2003.
Francis, Daniel. *A Road for Canada*. Vancouver: Stanton Atkins and Dosil, 2006.
Ganong, W.F. "The Vegetation of the Bay of Fundy Salt and Diked Marshes: An Ecological Study." *Botanical Gazette* 36 (1903): 161–86.
Giblett, Rodney James. *Canadian Wetlands: Places and People*. Bristol, UK: Intellect Books, 2014.
–. *Postmodern Wetlands: Culture, History, Ecology*. Edinburgh: Edinburgh University Press, 1996.
Gilbert, Jess. "Low Modernism and the Agrarian New Deal: A Different Kind of State." In *Fighting for the Farm*, edited by Jane Adams, 129–46. Philadelphia: University of Pennsylvania Press, 2001.
–. *Planning Democracy: Agrarian Intellectuals and the Intended New Deal*. New Haven, CT: Yale University Press, 2015.
Gray, James. *Men against the Desert*. Saskatoon: Western Producer Prairie Book, 1967.
Green, David F. "Chignecto Marshes: Bird and Hay Country." *Canadian Geographic* 103 (June/July 1981): 44–49.
Greenberg, David A., Wade Blanchard, Bruce Smith, and Elaine Barrow. "Climate Change, Mean Sea Level and High Tides in the Bay of Fundy." *Atmosphere-Ocean* 50, 3 (2012). https://doi.org/10.1080/07055900.2012.668670.
Harvey, Janice. "Death Watch on the Petitcodiac." *Atlantic Salmon Journal* 46 (1997): 36–43.
Hatvany, Matthew. "Imagining Duckland: Postnationalism, Waterfowl Migration, and Ecological Commons." *Canadian Geographer* 61, 2 (2017): 224–39.
–. *Marshlands: Four Centuries of Environmental Change on the Shores of the St. Lawrence*. Quebec: Presses de l'Université Laval, 2003.
–. "The Origins of the Acadian Aboiteau: An Environmental-Historical Geography of the Northeast." *Historical Geography* 30 (2002): 121–37.
–. "Wetlands and Reclamation." In *International Encyclopedia of Human Geography*, 2nd ed., edited by A. Kobayashi, 14: 273–79. Amsterdam: Elsevier, 2009. https://doi.org/10.1016/B978-008044910-4.00589-7.277.
Hatvany, Matthew, Donald Cayer, and Alain Parent. "Interpreting Salt Marsh Dynamics: Challenging Scientific Paradigms." *Annals of the Association of American Geographers* 105, 5 (2015): 1041–60.

Henton, Joseph, et al. "Crustal Motion and Deformation Monitoring of the Canadian Landmass." *Geomatica* 60, 2 (2006): 173–91.

Isaacman, Lisa. "Historic Examination of the Changes in Diadromous Fish Populations and Potential Anthropogenic Stressors in the Avon River Watershed, Nova Scotia." Master of Environmental Studies thesis, Dalhousie University, 2005.

Jackson, C.I., and J.W. Maxwell. *Landowners and Land Use in the Tantramar Area, New Brunswick*. CLI Report 9/Department of Energy, Mines and Resources, Geographical Paper No. 47. Ottawa: Canada Land Inventory and Department of Energy, Mines and Resources, 1971.

Josephson, Paul. *Industrialized Nature: Brute Force Technology and the Transformation of the Natural World*. Washington, DC: Island Press, 2002.

Judd, Richard. *Second Nature: An Environmental History of New England*. Amherst and Boston: University of Massachusetts Press, 2014.

Katopodis, Chris. *Introduction to Fishway Design*. Winnipeg: Freshwater Institute, 1991.

Kennedy, Gregory. "Marshland Colonization in Acadia and Poitou during the 17th Century." *Acadiensis* 42 (2013): 37–66.

–. *Something of a Peasant Paradise? Comparing Rural Societies in Acadie and the Loudanais, 1604–1755*. Montreal and Kingston: McGill-Queen's University Press, 2014.

Kenny, James L., and Andrew G. Secord. "Engineering Modernity: Hydroelectric Development in New Brunswick, 1945–70." *Acadiensis* 39 (2010): 3–26.

Kindervater, A.D. *Flooding Events in Nova Scotia: A Historical Perspective*. Halifax: Inland Water Directorate, Atlantic Region, 1977.

Langston, Nancy. *Where Water and Land Meet: A Western Perspective*. Seattle: University of Washington Press, 2005.

LeBlanc, Daniel. "Studies Show the Petitcodiac Will Largely Restore Itself." *Times & Transcript*, February 2, 2010.

LeBlanc, Ronnie-Gilles. "Documents acadiens sur les aboiteaux." *Les cahiers de la Société historique acadienne* 19 (1988): 39–48.

Leeming, Mark. *In Defence of Home Places: Environmental Activism in Nova Scotia*. Vancouver: UBC Press, 2017.

Leitch, W.G. *Ducks and Men*. Winnipeg: Ducks Unlimited Canada, 1978.

Lewey, Laurel, Louis J. Richard, and Linda Turner. *New Brunswick before the Equal Opportunity Program: History through a Social Work Lens*. Toronto: University of Toronto Press, 2018.

Locke, A., and R. Bernier. *Annotated Bibliography of Aquatic Biology and Habitat of the Petitcodiac River System, New Brunswick*. Moncton: Ocean Branch, Fisheries and Oceans Canada, 2000.

Longfellow, Henry Wadsworth. *Poems and Other Writings*. New York: Library of America, 2000.

Loo, Tina. "Disturbing the Peace: Environmental Change and the Scales of Justice on a Northern River." *Environmental History* 12 (2007): 895–919.

–. "High Modernism, Conflict, and the Nature of Change in Canada: A Look at *Seeing Like a State*." *Canadian Historical Review* 97 (2016): 34–58.

–. *Moved by the State: Forced Relocation and Making a Good Life in Postwar Canada*. Vancouver: UBC Press, 2019.

–. *States of Nature*. Vancouver: UBC Press, 2006.

Lynch, David K. "Tidal Bores." *Scientific American* 247, 4 (1982): 146–57.

Macfarlane, Daniel. *Negotiating a River: Canada, the US, and the Creation of the St. Lawrence Seaway.* Vancouver: UBC Press, 2014.

MacKinnon, Lachlan. "Five Simple Rules for Saving the Maritimes: The Regional Stereotype in the 21st Century." *Active History,* March 24, 2015, http://activehistory.ca/2015/03/five-simple-rules-for-saving-the-maritimes-the-regional-stereotype-in-the-21st-century.

MacLeod, Michaela. "The Causeway, Landfill, and the River: Shaping Moncton's Environs." Master of Architecture thesis, University of Waterloo, 2005.

Marchildon, Gregory. "The Prairie Farm Rehabilitation Administration: Climate Crisis and Federal-Provincial Relations during the Great Depression." *Canadian Historical Review* (90): 275–301.

Matas, Robert. "N.B. River Leads Endangered List." *Globe and Mail,* July 8, 2003. https://www.theglobeandmail.com/news/national/nb-river-leads-endangered-list/article 4128863.

Matthiessen, George. "Tidemarshes: A Vanishing Resource." In *Life in and around the Salt Marshes,* edited by Michael Ursin. New York: Thomas Cromwell, 1972.

McLaughlin, Mark. "Green Shoots: Aerial Insecticide Spraying and the Growth of Environmental Consciousness in New Brunswick, 1952–73." In *Land and Sea: Environmental History in Atlantic Canada,* edited by Claire Campbell and Robert Summerby-Murray, 143–57. Fredericton: Acadiensis Press, 2013.

Meindl, Christopher, ed. "Human Impacts on Wetlands." Special Section, *North American Geographer* 5 (2003): 154–232.

Mitsch, William J., and James G. Gosselink, eds. *Wetlands.* 4th ed. Hoboken, NJ: John Wiley and Sons, 2007.

Nicholas, George. "Putting Wetlands in Perspective," *Man in the Northeast* 42 (1991): 29–38.

Nomination Grand Pré. *World Heritage Nomination Proposal for the Landscape of Grand Pré.* Grand-Pré, NS, 2011.

Nova Scotia, Department of Agriculture and Marketing. *Maritime Dykelands: The 350 Year Struggle.* Halifax: Nova Scotia Department of Agriculture and Marketing, 1987.

Ouhilal, Yassine. "Flushed: Bore Culture, Civic Rescue, and the Legendary Tides of the Bay of Fundy." *Surfer's Journal* 23 (2014): 91–100.

Parker, J.S., L.W. McCarthy, and R.R. McIntyre. "Reclamation of Tidal Marshlands in the Maritime Provinces of Canada." Paper presented to the Soil and Water Group of the North Atlantic Section, American Society of Agricultural Engineers, August 27, 1951.

Parson, Helen E. "Regional Trends of Agricultural Restructuring in Canada." *Canadian Journal of Regional Science* 22, 3 (Fall 1999): 343–56.

Percy, J.A. "Dykes, Dams, and Dynamos." *Fundy Issues* 9 (1996): 1–6.

Pisani, Donald. *Water and American Government: The Reclamation Bureau, National Water Policy, and the West, 1902–35.* Berkeley: University of California Press, 2002.

Reid, Mark. "A River Divided." *Canada's History,* April-May 2010, 34–40.

Reilly, Nolan. "The General Strike in Amherst, Nova Scotia, 1919." *Acadiensis* 9 (1980): 56–77.

Reisner, Marc. *Cadillac Desert: The American West and Its Disappearing Water.* Revised ed. New York: Penguin, 1993.

Retson, G.C. "Marshland Farming in the Sackville Area of New Brunswick," *Canadian Farm Economics* 1 (1966): 20–26.

Rudin, Ronald. "The First Acadian Film: Visibility, Modernity, and Landscape in *Les aboiteaux*." *Canadian Historical Review* 96, 4 (2015): 507–33.
–. *Kouchibouguac: Removal, Resistance, and Remembrance at a Canadian National Park.* Toronto: University of Toronto Press, 2016.
–. *Remembering and Forgetting in Acadie: A Historian's Journey through Public Memory.* Toronto: University of Toronto Press, 2009.
Sandwell, R.W., ed. *Powering Up Canada: A History of Power, Fuel, and Energy from 1600.* Montreal and Kingston: McGill-Queen's University Press, 2016.
Scobie, Charlie. *People of the Tantramar*. Sackville, NB: Tantramar Heritage Trust, 2018.
Scott, James C. "High Modernist Social Engineering: The Case of the Tennessee Valley Authority." In *Experiencing the State*, edited by Lloyd I. Rudolph and John Kurt Jacobsen, 3–52. New Delhi: Oxford University Press, 2006.
–. *Seeing Like a State: How Certain Schemes to Improve the Human Condition Have Failed.* New Haven, CT: Yale University Press, 1998.
Sebold, Kimberly. "*The Low Green Prairies of the Sea*: Economic Usage and Cultural Construction of the Gulf of Maine Salt Marshes." PhD diss., University of Maine, 1998.
Spence, George. *Survival of a Vision*. Historical Series No. 3. Ottawa: Department of Agriculture, 1967.
Stunden Bower, Shannon. *Wet Prairie: People, Land and Water in Agricultural Manitoba.* Vancouver: UBC Press, 2011.
Summerby-Murray, Robert. "Commissioners of Sewers and the Intensification of Agriculture in the Tantramar Marshlands of New Brunswick." *North American Geographer* 5 (2003): 183–204.
–. "Interpreting Cultural Landscapes: A Historical Geography of Human Settlement on the Tantramar Marshes, New Brunswick." In *Themes and Issues in Canadian Geography III*, edited by Christoph Stadel, 157–74. Salzburg: Salzburger Geographische Arbeiten, 1999.
Szabo-Jones, Lisa. "Secrets of the Salt Marsh." *Alternatives Journal* 40 (2014): 24–31. https://www.alternativesjournal.ca/energy-and-resources/secrets-salt-marsh.
Teal, John, and Mildred Teal. *Life and Death of the Salt Marsh*. Boston: Little, Brown, 1969.
Thurston, Harry. *A Place between the Tides: A Naturalist's Reflections on the Salt Marsh.* Nanoose Bay, BC: Greystone Books, 2004.
–. *Tidal Life: A Natural History of the Bay of Fundy*. Willowdale, ON: Firefly Books, 1990.
Tiner, Ralph, ed. *Tidal Wetlands Primer*. Amherst and Boston: University of Massachusetts Press, 2013.
Tricker, R.A.R. *Bores, Breakers, Waves and Wakes: An Introduction to the Study of Waves on Water*. New York: American Elsevier, 1964.
Urquhart, M.C., ed., *Historical Statistics of Canada*. Cambridge/Toronto: Cambridge University Press/Macmillan, 1965.
Van Proosdij, Danika. *Dykelands: Climate Change Adaptation*. Halifax: Atlantic Climate Adaptation Solutions Association, Climate Change Directorate, Nova Scotia Department of Environment, 2011.
–. *Intertidal Morphodynamics of the Avon River Estuary*. Final report submitted to the Nova Scotia Department of Transportation and Public Works, September 30, 2007.
Van Proosdij, Danika, and Sarah M. Townsend, "Sedimentation and Mechanisms of Salt Marsh Colonization on the Windsor Mudflats, Minas Basin." In *The Changing Bay of*

Fundy: Beyond 400 Years. Proceedings of the 6th Bay of Fundy Workshop, Cornwallis, Nova Scotia – September 29th – October 2nd, 2004, edited by J.A. Percy et al., 258–79. Environment Canada, Atlantic Region, Occasional Report No. 23. Dartmouth, NS, and Sackville, NB: Environment Canada, 2005.

Vileisis, Ann. *Discovering the Unknown Landscape: A History of America's Wetlands.* Washington, DC: Island Press, 1997.

Wicken, William. "Re-examining Mi'kmaq-Acadian Relations, 1635–1755." In *Habitants et marchands, Twenty Years Later,* edited by Sylvie Dépatie et al., 93–114. Montreal and Kingston: McGill-Queen's University Press, 1998.

Williams, Michael, ed. *Wetlands: A Threatened Landscape.* Cambridge, MA: Basil Blackwood, 1991.

Worster, Donald. *Rivers of Empire: Water, Aridity, and the Growth of the American West.* New York: Oxford University Press, 1985.

Wynn, Graeme. "Late Eighteenth-Century Agriculture on the Bay of Fundy Marshlands." *Acadiensis* 8 (1979): 80–89.

Young, Robert A. "The Programme of Equal Opportunity: An Overview." In *The Robichaud Era, 1960–70: Colloquium Proceedings,* 23–35. Moncton: Canadian Institute for Research on Regional Development, 2001.

Zahra, Tara. "Lost Children: Displacement, Family, and Nation in Postwar Europe," *Journal of Modern History* 81 (2009): 45–86.

Index

Note: In subheadings, "MMRA" refers to "Maritime Marshland Rehabilitation Administration."

Acadians: beginnings on Île Ste-Croix, 9; construction of dykes and aboiteaux, 3, 9, 28–37; deportation, 9, 38–40; embrace of modernity, 8; opposition to Petitcodiac Causeway, 192–93, 197–98; use of aboiteaux as metaphor, 232n13; use of dykeland, 38
Amherst, Nova Scotia: site of MMRA headquarters, 77–78
Anderson, James, 127, 136
Anderson, Laurie: opposition to Tantramar Dam 117, 123, 134–36
Annapolis River Dam: and development along headpond, 147–48; and fish stocks, 144–45; and highway construction, 140–42; and pollution in headpond, 145–47
Archibald, E.S., 25, 238n105; and obstruction of tidal rivers, 118; and reducing number of aboiteaux, 55; and rehabilitation of dykelands, 58–59, 62–67; and tiding, 53–55
Avon Causeway: and accumulation of silt, 166, 208–12; and Windsor Salt Marsh, 208–12; as transportation project, 162; impact on fish stocks, 164–66, 212–15; public celebration of completion, 167–68; reconstruction of causeway, 214–15

Baird, W.W., 45–46; and Maritime Dykeland Rehabilitation Committee, 61, 64–65
Banks, C.A., 86–87, 130
Beaulieu, G.T., 160
Bishop Beckwith Marsh Body, 91
Black, A.E., 88
Bleakney, Conrad, 182–83
Boyer, George F. "Joe," 176–77
Buck, Bedford, 181–82, 191
Buckley, Helen: critique of MMRA, 109–11, 164
Butler, R.L., 144, 156, 159–60
Byers, J.W., 47, 50, 53, 63

Canadian Wildlife Service, 174–79
Castle Frederick Marsh Body, 73–74, 88
Centre Burlington Marsh Body, 97–98
Chadwick, Julia, 191, 205
Chiasson, Alyre, 196–98
Colas, Antony, 208

Index

Commissioners of Sewers, 40–41
Conlon, J.D., 81, 121, 131, 158, 162
cost-benefit ratio: as developed by MMRA, 93–98

Dentiballis Marsh Body, 42, 63
DesNeiges, Michel, 193–94, 200, 204
Desplanque, Constance "Con," 82–84
Dominion Experimental Farms, 25
Ducks Unlimited Canada, 177–79
dykelands: and deterioration of dykes and aboiteaux, 25–27; distribution of, 37; draining of, 18; and federal support for repairs, 58–67; and hay production, 42–46; introduction of heavy machinery, 48–50; in Kamouraska region of Quebec, 18–19; return to wetland, 174–79; and state intervention, 40–41; studied by experts, 46–51; unimproved after reconstruction work, 106–7, 173; and upland, 29–30; used for cattle grazing, 37
dykes and aboiteaux: construction of sluices, 36–37; deterioration, 4, 25–27; function of *le clapet*, 32, 37; and local knowledge, 32; sod cutting, 34–35

Elson, Paul, 156
Estabrooks, Ern, 137
Evangeline, 54, 238n110
expert knowledge, 11, 171; conflict with local knowledge, 11–12; conflicting forms of expert knowledge, 21; denigration of local knowledge, 50–51, 55–56, 74, 225; integration with local knowledge as métis, 12, 15–17, 56–57, 221, 225

Falmouth Great Dyke Marsh Body, 52, 73–74, 101, 222
Fawcett, W.W., 136
first nature, 18; destruction of, 28–37
Forest, Léonard: making *Les aboiteaux*, 3, 8, 10
Frail, George, 35, 81, 118–19

Ganong, W.F.: and tiding, 51, 54
Gardiner, J.L., 58, 68

Golding, William, 108–9
Graham, Shawn, 204
Grand Pre Marsh Body: collaboration with MMRA, 89–90; and creation of MMRA, 69–71; and introduction of heavy machinery, 48–50; meeting in 2019, 220–21, 223; and reconstruction work by MMRA, 100–1; settlement after deportation, 38–39; as UNESCO World Heritage Site, 10, 222–24
Griffin, Gary, 183, 187, 190–91, 205, 209, 217

Haase, Gordon, 49, 94–96, 99
Harkness, Douglas, 107
Harvey Bank Marsh Body, 61–62, 124
Hatfield, Richard, 187
Hazen, Douglas, 59, 69
Henry, C.E., 97
high modernism, 12–14, 22, 122, 225

Ilsley, J.L., 62–64

John Lusby Marsh Body, 76
John Lusby National Wildlife Area, 175

Kennedy, Robert F. Jr., 200
King, J.H., 136

Lake Pesaquid, 123, 214–15, 217
Lake Petitcodiac, 22, 123, 179–82; defended by residents, 188–91, 197–98, 203–4
Lake Petitcodiac Preservation Association (LAPPA), 179–80, 197–98, 206
Lattimer, J.E., 46
LeBlanc, Daniel, 193–94, 200–3, 206–7, 217
LeBlanc, Victor, 189
Lee, Sheldon, 181
Leenstra, Wiebe, 205
Les aboiteaux, 3, 6–8, 10–12, 32, 40, 48, 74
Lévesque, J. Adrien, 87, 103, 106–7
local knowledge, 5–6; denigration of, 50–51, 55–56, 74, 225; integration with expert knowledge as métis, 12, 15–17, 56–57, 221, 225; and upkeep of dykelands, 43

Log Lake Marsh Body, 137–38
Logan, Charles, 65–67
Logie, R.R., 145, 155–56
low modernism, 16

MacEachern, Neil, 165
Maritime Dykeland Rehabilitation Committee, 26–27, 60–66
Maritime Marshland Rehabilitation Administration (MMRA): accumulation of silt in rivers due to dam construction, 154–55; construction of tidal dams for transportation infrastructure, 19–21, 117–69, 148–51; creation of, 67–71; debate over value of its projects, 109–11; and decline of marsh bodies, 70–71, 104–5; as depicted in *Les Aboiteaux*, 7–8; development of engineering practices, 6–7, 79, 84–85, 90, 99; facilities in Amherst, Nova Scotia, 77–81; fish passage in tidal rivers, 144–45, 155–56; high-modernist practices, 168–69; integrated into Agricultural Rehabilitation and Development Administration (ARDA), 112, 248n142; integration of farmers' practices, 17, 74–75, 77, 79, 82–83; marsh owners held in low esteem by, 11, 14, 19, 68, 74, 87–88, 92–93, 121–22, 125–26; Marshland Utilization Committee, 93; obstruction of tidal rivers, 117–19, 121–23; pollution of headponds, 145–47; and process of simplification, 87–99; professional staff, 81–87; and provincial legislation, 69–70; and reconstruction of dykes and aboiteaux, 18–19, 99–101; return of projects to provinces, 105–8; and role of agricultural economists, 94; Special Committee on Drainage Problems, 90; and tiding, 121, 137–39; and use of photography, 85–87. *See also names of individual projects and officials*
marsh bodies, 12; decline due to New Brunswick legislation, 102–3; legislation creating, 40–41; role after creation of MMRA, 19, 69–71, 88–89, 101–10. *See also names of specific marsh bodies*

Marsh Owners' Association of Annapolis County, 26, 140
Matte, G.J., 163
McCarthy, L.W., 80, 84
McIntyre, Ron, 81, 84, 121, 127
McKenna, Frank, 187–92
McLaughlin, Victor, 179–81, 202
Mi'qmak First Nation, and marshland, 28–29
Munro, W.T., 176

Nappan Experimental Farm, 52, 61, 64
New Brunswick Marshland Reclamation Commission, 73, 134, 136
New Horton Marsh Body, 177
Niles, Eugene, 201–3, 205
Nova Scotia Marsh Survey, 46–47, 50, 53, 55
Nova Scotia Marshland Reclamation Commission, 97
Nova Scotia Royal Commission on Provincial Development and Rehabilitation, 26–27, 59

Palmeter family, 221–2; arrival in Grand-Pré, 38–39; Beverly, 223; Fred, 10; R.H., Chief Aboiteau Superintendent, 49, 82–84, 99–100; Robert, 221, 224
Palmeter-type aboiteau, 83
Parker, J.S., 68, 71, 74, 79, 81, 83–84, 89, 92–93, 96–100, 104–5, 225; appearance before Senate, 108–9; and construction of dams, 124–25, 131–38, 141–42; and Petitcodiac Causeway, 150–56; and pollution of headponds, 146–47
Partridge, Ernie, 3, 6–7, 35–36, 74, 80, 111–13, 224–25
Peterson, Robert, 75–76
Petitcodiac Causeway: and accumulation of silt, 154–55, 160–62, 183, 188–89, 207, 209; and creation of Lake Petitcodiac, 22; and degradation of river, 20–22; demands for opening gates, 186–95, 199–202, 217; and destruction of tidal bore, 151–54, 183, 207; and environmentalism, 171, 199–200; gates permanently opened, 204–8; as

high-modernist project, 149–50; and impact on fish stocks, 155–56, 158–60, 182–6, 188–89, 207; mobilization of youth against, 194–95; and Moncton's transportation needs, 148–51; and river pollution, 154, 156

Petitcodiac Lake Preservation Committee, 189

Planters: and settling on dykeland, 38–40

Prairie Farm Rehabilitation Administration, 16–17, 81

Program of Equal Opportunity (New Brunswick), 102–3

rehabilitation: of marshlands, 57–58; in postwar era, 13

Reicker, Jim, 204

Ritter, John, 189

Riverkeeper: Moncton chapter, 200; origins in US, 199–200

Roberts, J.A., 63–64, 73, 93, 103, 108, 127, 130, 132, 135, 141

Rogers, Ann, 225

Russell, Ben, 56–57, 140

salt marsh: draining of 28–37; value of production in, 22, 28

Scott, James C., 12, 89, 92, 132; and concept of mētis, 12, 15, 56, 77; and experts rejecting local knowledge, 55–56; and high modernism, 12–13;

second nature, 18–19, 37, 224; replaced by third nature, 217; restoration by MMRA, 67–68; restoration with opening of Petitcodiac Causeway gates, 205, 211; transformation into National Wildlife Areas, 178

Sellars, Jim, 203, 262n127

Semple, J.R., 183

Shepody National Wildlife Area, 177

Shepody River Dam, 120, 123–31

Smith, K.E.H., 164

Special Beef Cattle Committee, 45–48, 50

Spence-Sales, Harold, 149–50

surfing on Petitcodiac River, 153–5, 170, 172, 207–8

Taggart, J.G., 95–96

Tantramar Dam, 11, 117–21, 131–35; and construction of Trans-Canada Highway, 139; and efforts to secure popular support, 135–37

Taylor, Wilf, 159

third nature, 19, 119, 217; as created by Petitcodiac Causeway, 225; replaced by second nature with opening of Petitcodiac Causeway gates, 22; transformation of rivers, 20

Thompson, Toler, 131

tidal bore, 151–53

tidal dams: and accumulation of silt, 130; construction of, 19–20; as distinguished from aboiteaux, 119; and headponds, 130. *See also specific projects*

tiding: and construction of Tantramar Dam, 120–21, 131–35, 137–38; as long-standing practice, 51–53

Tihanyi, Eva: critique of MMRA, 109–11, 164

Tintamarre National Wildlife Area, 175–78

Trans-Canada Highway: and tidal dam construction, 139

water-diversion projects, 13–14

Waugh, John, 81–82, 162, 178–79

Wessels, JJ, 170

Whitbread, Colin, 170

Windsor Salt Marsh, 208–12, 215

Wood, Sonja, 214–15

NATURE | HISTORY | SOCIETY

Claire Elizabeth Campbell, *Shaped by the West Wind: Nature and History in Georgian Bay*
Tina Loo, *States of Nature: Conserving Canada's Wildlife in the Twentieth Century*
Jamie Benidickson, *The Culture of Flushing: A Social and Legal History of Sewage*
John Sandlos, *Hunters at the Margin: Native People and Wildlife Conservation in the Northwest Territories*
William J. Turkel, *The Archive of Place: Unearthing the Pasts of the Chilcotin Plateau*
Greg Gillespie, *Hunting for Empire: Narratives of Sport in Rupert's Land, 1840–70*
James Murton, *Creating a Modern Countryside: Liberalism and Land Resettlement in British Columbia*
Stephen J. Pyne, *Awful Splendour: A Fire History of Canada*
Sharon Wall, *The Nurture of Nature: Childhood, Antimodernism, and Ontario Summer Camps, 1920–55*
Hans M. Carlson, *Home Is the Hunter: The James Bay Cree and Their Land*
Joy Parr, *Sensing Changes: Technologies, Environments, and the Everyday, 1953–2003*
Liza Piper, *The Industrial Transformation of Subarctic Canada*
Jamie Linton, *What Is Water? The History of a Modern Abstraction*
Dean Bavington, *Managed Annihilation: An Unnatural History of the Newfoundland Cod Collapse*
J. Keri Cronin, *Manufacturing National Park Nature: Photography, Ecology, and the Wilderness Industry of Jasper*
Shannon Stunden Bower, *Wet Prairie: People, Land, and Water in Agricultural Manitoba*
Jocelyn Thorpe, *Temagami's Tangled Wild: Race, Gender, and the Making of Canadian Nature*
Sean Kheraj, *Inventing Stanley Park: An Environmental History*
Darcy Ingram, *Wildlife, Conservation, and Conflict in Quebec, 1840–1914*
Caroline Desbiens, *Power from the North: Territory, Identity, and the Culture of Hydroelectricity in Quebec*
Daniel Macfarlane, *Negotiating a River: Canada, the US, and the Creation of the St. Lawrence Seaway*

Justin Page, *Tracking the Great Bear: How Environmentalists Recreated British Columbia's Coastal Rainforest*

Ryan O'Connor, *The First Green Wave: Pollution Probe and the Origins of Environmental Activism in Ontario*

John Thistle, *Resettling the Range: Animals, Ecologies, and Human Communities in British Columbia*

Jessica van Horssen, *A Town Called Asbestos: Environmental Contamination, Health, and Resilience in a Resource Community*

Nancy B. Bouchier and Ken Cruikshank, *The People and the Bay: A Social and Environmental History of Hamilton Harbour*

Carly A. Dokis, *Where the Rivers Meet: Pipelines, Participatory Resource Management, and Aboriginal-State Relations in the Northwest Territories*

Jonathan Peyton, *Unbuilt Environments: Tracing Postwar Development in Northwest British Columbia*

Mark R. Leeming, *In Defence of Home Places: Environmental Activism in Nova Scotia*

Jim Clifford, *West Ham and the River Lea: A Social and Environmental History of London's Industrialized Marshland, 1839–1914*

Michèle Dagenais, *Montreal, City of Water: An Environmental History*

David Calverley, *Who Controls the Hunt? First Nations, Treaty Rights, and Wildlife Conservation in Ontario, 1783–1939*

Jamie Benidickson, *Levelling the Lake: Transboundary Resource Management in the Lake of the Woods Watershed*

Daniel Macfarlane, *Fixing Niagara Falls: Environment, Energy, and Engineers at the World's Most Famous Waterfall*

Angela V. Carter, *Fossilized: Environmental Policy in Canada's Petro-Provinces*

Stéphane Castonguay, *The Government of Natural Resources: Science, Territory, and State Power in Quebec, 1867–1939*

Printed and bound in Canada by Friesens
Set in Garamond by Artegraphica Design Co. Ltd.
Copy editor: Frank Chow
Proofreader: Caitlin Gordon-Walker
Cartographer: Eric Leinberger
Cover designer: Setareh Ashrafologhalai

Cover images: *top left:* Trying to hold back the tides. Les aboiteaux. *Copyright 1955 National Film Board of Canada. All rights reserved. Collection: Cinémathèque québécoise; top right:* Hayfield in Tantramar Marsh. *Mount Allison University Archives, Robert J. Cunningham Fonds, 7922/4b/29; bottom right:* Surfers on the Petitcodiac, July 2013. *Courtesy of Trevor Gertridge; bottom left:* Rebuilding an aboiteau. Les aboiteaux. *Copyright 1955 National Film Board of Canada. All rights reserved. Collection: Cinémathèque québécoise.*